PION–NUCLEON SCATTERING

PION–NUCLEON SCATTERING

EDITED BY

GORDON L. SHAW

University of California
Irvine, California

AND

DAVID Y. WONG

University of California
San Diego, California

Wiley-Interscience

A DIVISION OF JOHN WILEY & SONS, NEW YORK · LONDON · SYDNEY · TORONTO

Library of Congress Catalog Card Number 69-18014
SBN 471 779954

Printed in the United States of America

Foreword

A πN Scattering Conference was held December 1 and 2, 1967 at the University of California, Irvine. The Conference, sponsored by the U.S. Atomic Energy Commission and UCI, was organized by G. F. Chew, G. L. Shaw, and D. Y. Wong.

The first sessions consisted of contributed papers. This was followed by nine summary talks and a panel discussion. The panel members were G. F. Chew, R. H. Dalitz, C. Lovelace, and J. J. Sakurai; the chairman was S. Mandelstam. In this volume, only the nine summary talks and the more lengthy parts of the panel discussion by Chew, Dalitz, and Sakurai are included.

Participation in the Conference was open to all interested physicists and about 120 attended. This was a small enough number so that a good deal of informality was present and there was much discussion back and forth between the speakers and the audience. There was a general feeling that the concept of a specialized, open-participation conference of short duration is a good one.

The introduction includes a numerical table of πN kinematics and a number of relations relevant to πN scattering. We hope that this volume will be useful not only to workers in hadron physics, but also to graduate students.

G. L. SHAW
D. Y. WONG

Introduction

The scattering of pions by nucleons has played a particularly significant role in the development of particle physics. On the one hand, it is readily accessible to experimental investigations because of the availability of high intensity pion beams over a wide range of energies. On the other hand, it provides a testing ground for many of the theoretical ideas. The well-known (3, 3) resonance $N^*(1238)$ has yielded a test of charge independence for strong interactions, verified the formalism of final state interactions relating πN scattering to photopion production, and stimulated the development of the Chew-Low theory. The forward scattering of pions by nucleons has served as the basis for testing the validity of microcausality through dispersion relations. Many new resonances have been found in the phase shift analyses and these results have imposed severe constraints on dynamical models such as the composite particle models of the nucleon and the N^*'s. The high energy πN scattering amplitudes in both the forward and the backward directions are essential in establishing the correctness of the Regge pole-dominant theory. The S-wave scattering lengths have provided a crucial check of the self-consistency of the soft-pion theory. Analysis now in progress of inelastic events at moderate energies holds the promise of many interesting results. It is rather remarkable that these many important comparisons between theory and experiment can be made within the realm of πN scattering. Although a proper perspective of strong interaction physics must be acquired through the study of many hadronic systems, nevertheless, it is possible to obtain useful and informative results through the investigation of the πN system alone.

A collection of formulas relevant to πN scattering is given below. We note that the notation and normalizations used by various authors (e.g., in this volume) is not uniform.

Let the pion four-momentum be denoted by q and the nucleon four-momentum by p. The elastic scattering process is

$$\pi_1 + N_1 \rightarrow \pi_2 + N_2 \tag{1}$$

with

$$q_1 + p_1 = q_2 + p_2$$

The Lorentz scalar variables are

$$s = (p_1 + q_1)^2$$
$$t = (q_1 - q_2)^2 \tag{2}$$
$$u = (p_1 - q_2)^2$$

with

$$s + t + u = 2m^2 + 2\mu^2$$

where m and μ are the masses of the nucleon and pion, respectively. The scalar s is physically interpreted as the energy variable, t the momentum transfer variable (and u the momentum transfer from the pion to the nucleon).

In the center-of-mass system, let W be the total energy, k the momentum, and θ the scattering angle. They are related to s and t by

$$W = s^{1/2}$$
$$k^2 = (s - (m + \mu)^2)(s - (m - \mu)^2)/4s \tag{3}$$
$$t = -2k^2(1 - \cos\theta)$$

In the laboratory system, the incident meson total energy v is given by

$$v = (s - m^2 - \mu^2)/2m \tag{4}$$

In Table I, we present some πN kinematics.

TABLE I

Laboratory Kinetic Energy of Incident Pion T, Center-of-Mass Energy W, and Center-of-Mass Momentum k versus Incident Pion Laboratory Momentum P_L. The Masses μ and m are taken as 140 MeV and 938 MeV, respectively.

P_L(MeV/c)	T(MeV)	W(MeV)	k(MeV/c)
20	1.4	1079	17
40	5.6	1083	35
60	12	1089	52
80	21	1096	68
100	32	1106	85
120	44	1116	101
140	58	1127	116
160	73	1139	132
180	88	1152	147
200	104	1165	161

(continued)

TABLE I (*continued*)

P_L(MeV/c)	T(MeV)	W(MeV)	k(MeV/c)
250	147	1199	196
300	191	1233	228
350	237	1268	259
400	284	1302	288
450	331	1335	316
500	379	1369	343
550	428	1401	368
600	476	1434	393
650	525	1465	416
700	574	1496	439
750	623	1527	461
800	672	1557	482
850	721	1586	503
900	771	1615	523
950	820	1643	542
1000	870	1671	561
1050	919	1699	580
1100	969	1726	598
1150	1018	1753	615
1200	1068	1779	633
1250	1118	1805	649
1300	1168	1831	666
1350	1217	1856	682
1400	1267	1881	698
1450	1317	1906	714
1500	1367	1930	729
1550	1416	1954	744
1600	1466	1978	759
1650	1516	2001	773
1700	1566	2025	788
1750	1616	2048	802
1800	1665	2070	816
1850	1715	2093	829
1900	1765	2115	843
1950	1815	2137	856
2000	1865	2159	869
2200	2064	2244	920
2400	2264	2366	968
2600	2464	2405	1014
2800	2663	2482	1058
3000	2863	2556	1101

(*continued*)

TABLE I (*continued*)

P_L(MeV/c)	T(MeV)	W(MeV)	k(MeV/c)
3200	3063	2628	1142
3400	3263	2699	1182
3600	3463	2767	1220
3800	3663	2834	1258
4000	3862	2900	1294
5000	4862	3207	1463
6000	5862	3487	1614
7000	6861	3746	1753
8000	7861	3989	1881
9000	8861	4217	2002
10000	9861	4434	2115
11000	10861	4641	2223
12000	11861	4839	2326
13000	12861	5029	2425
14000	13861	5212	2520
15000	14861	5389	2611
16000	15861	5560	2699
17000	16861	5726	2785
18000	17861	5888	2868
19000	18861	6045	2948
20000	19860	6198	3027
21000	20860	6348	3103
22000	21860	6494	3178
23000	22860	6637	3251
24000	23860	6777	3322
25000	24860	6914	3392
26000	25860	7048	3460
27000	26860	7180	3527
28000	27860	7309	3593
29000	28860	7437	3658
30000	29860	7562	3721
31000	30860	7685	3784
32000	31860	7806	3845
33000	32860	7925	3906
34000	33860	8043	3965
35000	34860	8158	4024
36000	35860	8273	4082
37000	36860	8385	4139
38000	37860	8496	4195
39000	38860	8606	4251
40000	39860	8714	4306

All charge combinations of the scattering process (1) are expressed in terms of isotopic spin $I = \frac{1}{2}$ and $\frac{3}{2}$ amplitudes. The amplitudes in the expressions below for the cross sections will be the appropriate sum, whereas unless explicitly mentioned the other relations hold separately for each I.

The differential cross section can be expressed in terms of two amplitudes $f_1(s, t)$ and $f_2(s, t)$ as

$$\frac{d\sigma}{d\Omega} = \sum_{i,f} \left| \chi_f^+ \left[f_1(s, t) + \frac{(\boldsymbol{\sigma} \cdot \mathbf{q}_1)(\boldsymbol{\sigma} \cdot \mathbf{q}_2)}{|\mathbf{q}_1| \, |\mathbf{q}_2|} f_2(s, t) \right] \chi_i \right|^2 \tag{5}$$

where the χ's are the two-component spin wavefunctions and Σ denotes a sum over the final spin state f and an average over the initial state i appropriate to the experimental arrangement. The partial-wave expansions of f_1 and f_2 in the center-of-mass system are given by

$$f_1 = \sum_{l=0}^{\infty} f_{l+} P'_{l+1}(\cos \theta) - \sum_{l=2}^{\infty} f_{l-} P'_{l-1}(\cos \theta)$$

$$f_2 = \sum_{l=1}^{\infty} (f_{l-} - f_{l+}) P'_l(\cos \theta) \tag{6}$$

where $P'_{l-1} = dP_l/d \cos \theta$ and the subscripts $l\pm$ correspond $J = l \pm \frac{1}{2}$. The normalization of the partial-wave amplitude f_J is

$$f_J = (S_J - 1)/2ik$$

$$S_J = \eta e^{2i\delta} \tag{7}$$

where δ is real and the inelastic factor η takes on values $0 \le \eta \le 1$. Thus in the physical scattering region, the unitarity relation is

$$\mathrm{Im}\, f_J = k \, |f_J|^2 + (1 - \eta^2)/4k \tag{8}$$

Frequently, a partial-wave amplitude is denoted by $l_{2I,2J}$, e.g., P_{33}.

The expression connecting the differential cross section to the helicity amplitudes $f_{\lambda_f \lambda_i}$ is somewhat simpler:

$$\frac{d\sigma}{d\Omega} = \sum_{i,f} |f_{\lambda_i \lambda_f}|^2 \tag{9}$$

Using $\pm$ for the nucleon helicity, one obtains

$$f_{++} = f_{--} = (f_1 + f_2) \cos \frac{\theta}{2}$$

$$f_{+-} = -f_{-+}^* = (f_1 - f_2) \sin \frac{\theta}{2} e^{-i\phi} \tag{10}$$

Both sets $\{f_1, f_2\}$ and $\{f_{\lambda_f}, f_{\lambda_i}\}$ have kinematical singularities. They are related to kinematic singularity free Lorentz invariant amplitudes A and B by

$$f_1 = (E + m)/(8\pi W)[A + (W - m)B]$$
$$f_2 = (E - m)/(8\pi W)[-A + (W + m)B] \qquad (11)$$

where the energy of the nucleon

$$E = (s + m^2 - \mu^2)/2W$$

The amplitudes A and B are related to the T matrix by

$$T = -A + \tfrac{1}{2}\gamma \cdot (q_1 + q_2)B \qquad (12)$$

This T in turn is related to the cross section by

$$\frac{d\sigma}{d\Omega} = \sum \left| \bar{u}_f \frac{m}{4\pi W} T u_i \right|^2 \qquad (13)$$

where the u's are the four-component Dirac spinors normalized to $\bar{u}u = 1$.

In terms of partial-wave projections of A and B,

$$\{A_l(s), B_l(s)\} \equiv \int_{-1}^{1} d\cos\theta P_l(\cos\theta)\{A(s, t), B(s, t)\} \qquad (14)$$

we have

$$f_{l\pm}(W) = (1/16\pi W)\{(E + m)[A_l + (W - m)B_l]$$
$$+ (E - m)[-A_{l\pm 1} + (W + m)B_{l\pm 1}]\} \qquad (15)$$

From (15) we observe the MacDowell symmetry relation

$$f_{l+}(W) = -f_{(l+1)-}(-W) \qquad (16)$$

The amplitudes A and B not only describe the elastic scattering process (1) but also the crossed processes

$$\bar{\pi}_2 + N_1 \rightarrow \bar{\pi}_1 + N_2 \qquad (17)$$

and

$$\pi_1 + \bar{\pi}_2 \rightarrow \bar{N}_1 + N_2 \qquad (18)$$

Processes (1), (17), and (18) are called the s, u, and t channel reaction, respectively, since in the center-of-mass system for each channel, the named scalar is the square of the total energy. It is usually assumed that A and B obey the Mandelstam representation, which allows one to analytically continue from the physical region of one channel to another. The position of the singularities of f_J in the W plane are shown in Fig. 1. The t channel $(\pi\pi \rightarrow N\bar{N})$ contributes to the circle of radius $(m^2 - \mu^2)^{1/2}$ centered at

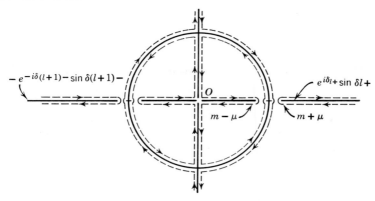

$$-e^{-i\delta(l+1)} - \sin \delta(l+1) - \qquad e^{i\delta_l} + \sin \delta l +$$

FIG. 1. The position of the singularities in the W plane of f_J. The branch cuts are indicated by heavy lines; the dotted line is the contour of integration for the partial-wave dispersion relation. Figures 1 and 2 are taken from W. Frazer and J. Fulco, *Phys. Rev.*, **119**, 1420 (1960).

$W = 0$ as well as a cut along the imaginary axis. The branch cuts along the real axis in the s-plane arising from the u channel pole at m^2 and continuum starting $(m + \mu)^2$ are given in Fig. 2. The pole at $u = m^2$ gives the short cut $m^2 - 2\mu^2 + \mu^4/m^2 \le s \le m^2 + 2\mu^2$ and one for $s < 0$.

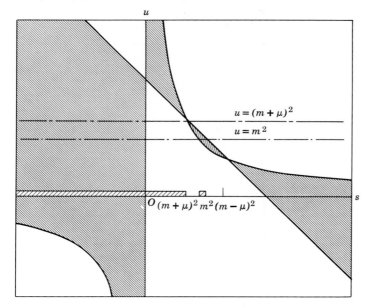

FIG. 2. The real su plane for πN scattering. The shaded area is the region in which $1 \le \cos \theta \le 1$. The intersection of this area with the region of u in the Mandelstam representation $[u = m^2, u \ge (m + \mu)^2]$ gives rise to the branch cuts shown along the s axis.

The nucleon pole terms contribute to the B amplitude only. They are given by

$$B^{I=1/2}(s, t) = 3g^2/(m^2 - s) + g^2/(m^2 - u)$$
$$B^{I=3/2}(s, t) = -[2g^2/(m^2 - u)]$$

(19)

with $g^2/4\pi \simeq 14$.

The invariant amplitudes are not only free of kinematical singularities, but they also have a simple symmetry property under the interchange of s and u (crossing symmetry). Consider the following combination of amplitudes with $I = \frac{1}{2}$ and $\frac{3}{2}$:

$$A^{(+)} = \tfrac{1}{3}(A^{(1/2)} + 2A^{(3/2)})$$
$$A^{(-)} = \tfrac{1}{3}(A^{(1/2)} - A^{(3/2)})$$

(20)

and similarly for the B amplitudes. Crossing symmetry implies that

$$A^{(\pm)}(s, t) = \pm A^{(\pm)}(u, t)$$
$$B^{(\pm)}(s, t) = \mp B^{(\pm)}(u, t)$$

(21)

The ($\pm$) amplitudes are proportional to the $I = 0, 1$ $\pi\pi \to N\bar{N}$ amplitudes with invariant energy t and momentum transfer s, and the above symmetry is the symmetry with respect to the interchange of the two pions. Explicitly, $A^{(0)} = \sqrt{6}\, A^{(+)}$, $A^{(1)} = 2A^{(-)}$ (similarly for B) and the relations between the $\pi\pi \to N\bar{N}$ helicity amplitudes and A and B are

$$f^{(0,1)}_{++,00}(t, s) = -\frac{1}{8\pi}\left(\frac{t - 4m^2}{t - 4\mu^2}\right)^{1/4}\frac{1}{[t(t - 4m^2)]^{1/2}}[(t - 4m^2)A^{(0,1)}(s, t)$$
$$+ (2s + t - 2m^2 - 2\mu^2)B^{(0,1)}(s, t)]$$

(22)

$$f^{(0,1)}_{+-,00}(t, s) = \frac{1}{16\pi}\left(\frac{t - 4m^2}{t - 4\mu^2}\right)^{1/4}(t - 4\mu^2)^{1/2}B^{(0,1)}(s, t)e^{i\varphi}$$

The differential cross section for $\pi\pi \to N\bar{N}$ is given by

$$\frac{d\sigma}{d\Omega} = \sum |f_{\lambda_1\lambda_2,00}|^2$$

In the Regge pole analyses, it is useful to express the πN scattering cross section also in terms of the $\pi\pi \to N\bar{N}$ amplitudes. One finds for the spin-average differential cross section

$$\left(\frac{d\sigma}{d\Omega}\right)_{\pi N \to \pi N} = \frac{t}{s}\left(\frac{t - 4\mu^2}{t - 4m^2}\right)^{1/2}(|f_{++,00}|^2 + |f_{+-,00}|^2)$$

(23)

Note that the forward cross section only involves $f_{++,00}$ which has a $1/t^{1/2}$ singularity ($f_{+-,00}$ is finite). The polarization P is also simply related to the t-channel as well as the s-channel helicity amplitudes. The formula is

$$\left[P\left(\frac{d\sigma}{d\Omega}\right)\right]_{\pi N \to \pi N} = 2 \operatorname{Im} f_{+-}{}^* f_{++}$$

$$= \frac{2t}{s}\left(\frac{t-4\mu^2}{t-4m^2}\right)^{1/2} \operatorname{Im} f_{+-,00}^* f_{++,00} \tag{24}$$

G .L. Shaw

D. Y. Wong

Contributors

GEOFFREY F. CHEW, *University of California, Berkeley, California*

R. H. DALITZ, *Department of Theoretical Physics, Oxford University, Oxford, England*

STEVEN FRAUTSCHI, *California Institute of Technology, Pasadena, California*

CLAIBORNE JOHNSON, *Lawrence Radiation Laboratory, University of California, Berkeley, California*

S. J. LINDENBAUM, *Brookhaven National Laboratory, Upton, New York*

C. LOVELACE, *CERN, Geneva Switzerland.*

STANLEY MANDELSTAM, *Department of Physics, University of California, Berkeley, California*

ARTHUR H. ROSENFELD, *Department of Physics and Lawrence Radiation Laboratory, University of California, Berkeley, California*

MARC ROSS, *University of Michigan, Ann Arbor, Michigan*

J. J. SAKURAI, *The Enrico Fermi Institute for Nuclear Studies and the Department of Physics, The University of Chicago, Chicago, Illinois*

HOWARD J. SCHNITZER, *Department of Physics, Brandeis University, Waltham, Massachusetts*

GORDON L. SHAW, *Department of Physics, University of California, Irvine, California*

PAUL H. SÖDING, *Department of Physics and Lawrence Radiation Laboratory, University of California, Berkeley, California*

HERBERT STEINER, *Lawrence Radiation Laboratory, University of California, Berkeley, California*

Contents

PION–NUCLEON SCATTERING

πp Elastic Scattering: Low Energy Experiments and Phase Shift Analyses

CLAIBORNE JOHNSON and HERBERT STEINER

Lawrence Radiation Laboratory,
University of California, Berkeley, California

I. Introduction

It is the purpose of this paper to review briefly the experimental situation and the results of the various phase shift analyses, confining ourselves to $p_\pi \lesssim 2$ GeV/c. No startling new experimental results have been reported during the last year, although some further cross section and polarization measurements have been completed. On the other hand, the phase shifters have been busy—so busy in fact that the number of reported πN resonances doubled during 1967.

What we would like to do then is first to bring you up to date on the present status of πP elastic scattering experiments below $p_\pi \sim 2$ GeV/c. Then we will discuss the phase shift situation—the methods used, the results obtained, and some of the problems encountered by the various groups engaged in this type of work.

II. Experiments

A. Possible Experiments

All of the experiments that can possibly be done by elastically scattering pions on nucleons can be conveniently summarized by the equation:

$$I\langle\sigma_\mu\rangle_f = I_0 \sum_{v=0}^{3} D_{\mu v}\langle\sigma_v\rangle_i$$

where

I = the scattered intensity
I_0 = the scattered intensity when the initial state nucleon is unpolarized
$\sigma_\mu = (\sigma_0, \sigma_1, \sigma_2, \sigma_3)$ where σ_0 is the (2×2) unit matrix, 1, and $\sigma_1, \sigma_2, \sigma_3$ are the three Pauli spin matrices

1

the subscripts f and i refer to the final and initial states, respectively. $D_{\mu\nu}$ is the depolarization operator which acts on the initial state of polarization to produce the final polarization state. For example, $D_{j0}\ j = 1, 2, 3$ refers to that experiment in which the 3 components of the vector polarization of the nucleon are determined when the initial state is unpolarized. Similarly D_{kl} would describe the experiment where the target proton is polarized along the l direction, and measurement is made of the final proton's polarization in the k direction.

In terms of the M-matrix which acts on the wavefunction describing the initial state to produce the final state wavefunction, D can be written

$$D_{\mu\nu} = \tfrac{1}{2}\,\mathrm{Tr}(M\sigma_\nu M^\dagger \sigma_\mu)/\tfrac{1}{2}\,\mathrm{Tr}(MM^+)$$

An explicit representation for D can be obtained if one chooses a specific form for the M-matrix. For example, let us use the often-used form

$$M = G1 + iH\boldsymbol{\sigma}\cdot\hat{n}$$

where G and H are functions of c.m. energy and angle, **1** is the (2×2) unit matrix, $\boldsymbol{\sigma}\cdot\hat{n}$ is the component of the spin operator in the direction normal to the scattering plane; i.e., $\hat{n} = \hat{k}_i \times \hat{k}_f$. $\hat{k}_i$ is the direction of the incident nucleon in the center of mass (along the $+z$ axis). $\hat{k}_f$ is the direction of the final nucleon in the center of mass (in x–z plane at angle θ with respect to the z axis). Then D can be written:

f \ i	0	y	z	x
0	1	α	0	0
y	α	1	0	0
z	0	0	γ	β
x	0	0	$-\beta$	γ

$D =$

where

$$I_0\alpha = 2\mathrm{Im}\ GH^*$$
$$I_0\beta = 2\mathrm{Re}\ GH^*$$
$$I_0\gamma = |G|^2 - |H|^2$$
$$I_0 = |G|^2 + |H|^2$$

The z–x plane is the scattering plane, y is the normal to the scattering plane. There are no elements of D connecting the $(0, y)$ components with the (z, x) components. This is a consequence of parity conservation in the scattering process. (I chose the rather odd order of labeling the elements of the matrix written above to show its block diagonal nature when parity is conserved.) Note also that $\gamma = 1$ means that there is no spin-flip whereas $\gamma = -1$ implies that there is only spin-flip.

The various elements of the D-matrix can be directly related to the so-called Wolfenstein parameters.[1] For example,

$$P = D_{0y} = D_{y0} = \alpha$$
$$D = D_{yy} = 1$$

A and R are related to β and γ as will be shown below. From the experimental point of view every time you see a nonzero element in any of the last three columns it means that that experiment involves the use of a polarized target. Every time you see a nonzero element in any of the last three rows it means that the experiment involves an additional scattering to analyze the polarization of the final state nucleon. Thus only P and I_0 can be determined by experiments involving a single scattering. This is the main reason that most of the experimental effort up to now has been devoted to measurements of these parameters. One slight complication arises due to the fact that most high energy experiments up to now at least have been done not in the center of mass but in the laboratory system. The polarization components normal to the scattering plane are unchanged under transformation from center-of-mass to laboratory frame of reference. On the other hand, care must be taken in relating measurements in the lab involving polarization components in the plane of the scattering to the components of the depolarization tensor which is defined in the center of mass. In 1954 Wolfenstein[1] introduced the parameters A and R to describe the change of polarization in the plane of the scattering of the incident particle in the lab. In πN scattering where the incident projectile has spin zero it is convenient to introduce analogous parameters A_{recoil} and R_{recoil} which refer instead to the change of the target nucleon's polarization in the plane of the scattering. (See Fig. 1.)

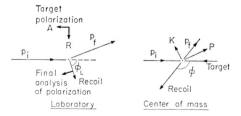

FIG. 1. Geometry for measurement of depolarization parameters, R and A, using a polarized target.

$$A_{\text{recoil}} = -\beta \cos(\phi - \phi_L) + \gamma \sin(\phi - \phi_L)$$
$$R_{\text{recoil}} = +\beta \sin(\phi - \phi_L) + \gamma \cos(\phi - \phi_L)$$

Up to now only experiments involving measurements of total cross sections, differential cross sections, and polarization normal to the scattering plane have been reported. However, A_{recoil} and R_{recoil} measurements are presently underway at CERN by a group from Saclay. These measurements are being made at energies higher than those considered in this report, but it seems likely that within the next few years we can expect to obtain information on β and γ also at lower energies. It should be kept in mind that A- and R-type measurements involve both the use of a polarized target and the subsequent analysis of the recoil proton's polarization. Consequently experimental complications are likely to limit the accuracy and scope of these measurements for a while at least.

B. New Results 1966–1967

There have been five experiments reported during the last year. A group at University College and Westfield College, London, has measured $d\sigma/dt$ for $\pi^- p$ scattering at 5 momenta between 1.72 and 2.46 GeV/c,[2] and for $\pi^+ p$ scattering at 10 momenta between 1.72 and 2.80 GeV/c.[3] In Fig. 2 we show some the these results. The purpose in showing these results here is to give you some idea of the quality of the data and the angular range covered. These results are typical or perhaps even slightly better than other measurements of the same type reported by other groups in the past at different energies.

A group at Brookhaven has reported[4] measurements of differential cross sections in backward directions for both $\pi^+ p$ and $\pi^- p$ elastic scattering. These results were presented in one of the contributed papers to this conference.

New measurements of $\pi^+ p$ and $\pi^- p$ total cross sections for momenta between 0.5 and 2.65 GeV/c have recently been reported. These results, which are of very high quality, were obtained in a collaborative effort between Cambridge, Rutherford, and Birmingham at Nimrod.[5]

Another group,[6] also at Nimrod, has made very detailed measurements of polarization in $\pi^- p$ scattering at 50 different momenta between 0.64 and 2.14 GeV/c. As you will see, they did a very thorough job of it.

The last experiment I want to mention is also a measurement of the polarization in $\pi^\pm p$ scattering at 364, 425, 490, and 532 MeV/c at the 184-in. synchrocyclotron at Berkeley.[7] These results are not in final form but are of interest nevertheless because they disagree in places with the results of similar experiments reported earlier. In particular the polarizations in $\pi^+ p$ scattering at 532 and π^- at 427 MeV/c are shown in Figs. 3 and

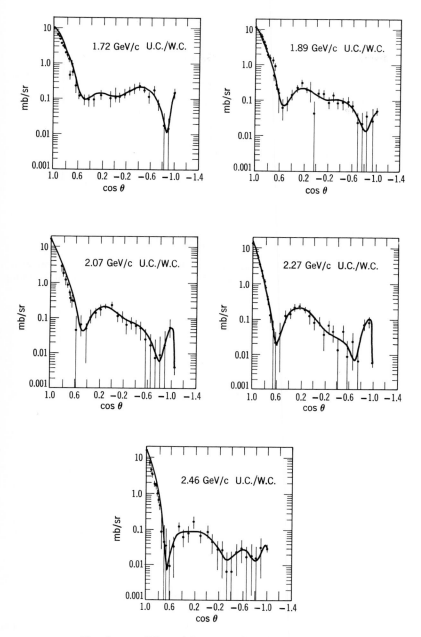

FIG. 2. $\pi^- p$ differential cross sections from Ref. 2.

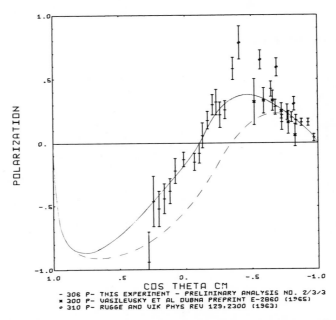

FIG. 3. Polarization in $\pi^- p$ scattering at 306 MeV as reported by Gorn et al.[7] The dashed and solid curves are phase-shift fits obtained before including and after including the new data.

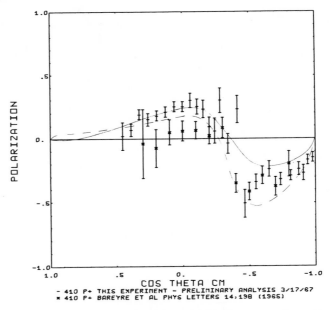

FIG. 4. Polarization in $\pi^+ p$ scattering at 410 MeV as reported by Gorn et al.[7] The dashed and solid curves are phase-shift fits obtained before including and after including the new data.

4. Also shown are the results of previous experiments and the phase shift fits. It is interesting to point out that although the data have changed, the phase shifts needed to fit the new results differ only very minimally from those which fit the earlier data.

C. Summary of Current Status of πp Scattering Experiments

In Figs, 5. and 6 we have listed the momenta at which various measurements have been made. It is seen that the differential cross sections for π^+ and π^- elastic scattering are in pretty good shape, although I would add that experiments of higher accuracy would be very welcome indeed. This last statement applies even more to differential cross-section data for the process $\pi^- p \to \pi^0 n$. The results here are not as reliable as one could wish and further measurements would be very useful, especially at momenta of 1180, 1360, and 1440 MeV/c.

With respect to polarization measurements the greatest need obviously lies in $\pi^- p$ charge exchange. In the energy region considered here a total of one measurement, at one angle at one energy (310 MeV) exists,[8] and even it is of rather low statistical accuracy. It is worth noting that charge-exchange polarization results in the momentum region $2 \le p\pi \le 5.5$ GeV/c have recently been reported by a group from Argonne.[9] These measurements, though far from easy, are feasible and are likely to become more extensive in the near future. I think most of the phase shifters would agree that this type of experiment is probably of greatest interest at present. The

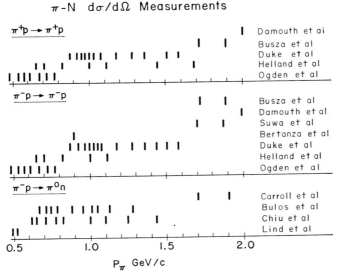

FIGURE 5.

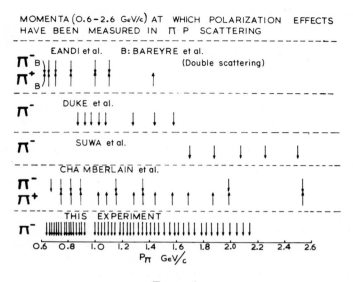

FIGURE 6.

polarization parameter in $\pi^- p$ elastic scattering has been very thoroughly measured—especially by the Nimrod Group[6]—so that for the time being at least no further measurements are called for. Polarization in $\pi^+ p$ scattering, on the other hand, has not been investigated as exhaustively and more detailed measurements would be welcome. To give you an idea of the quality of the existing polarization data, the results of the Berkeley group[10] at $p_\pi = 1.352$ GeV/c are shown in Fig. 7. These are typical of the

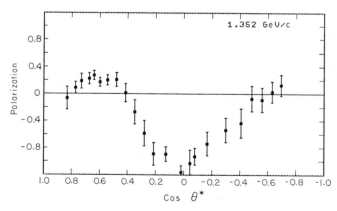

FIG. 7. Polarization in $\pi^- p$ scattering at $p_\pi = 1.352$ GeV/c as reported by Chamberlain et al.[10]

results obtained by other groups at different energies. A complete summary of references of the published πp experiments may be found in Ref. 11.

It might be worthwhile to try to see if there are any trends or patterns to be observed in looking at these data. There are significant forward peaks in $d\sigma/dt$ for $\pi^+ p \to \pi^+ p$, $\pi^- p \to \pi^- p$, and even in $\pi^- p \to \pi^0 n$. Generally speaking the first minimum for $\pi^- p$ elastic scattering is at $t \simeq -0.5$ GeV^2 whereas the corresponding minimum for $\pi^+ p$ elastic scattering is somewhat wider, say $t \simeq -0.6$ to -0.8 GeV^2. The polarization results near $t = 0$ usually show $P_{\pi^+} > 0$, whereas P_{π^-} is most often negative. The first zero in the polarization parameter for both π^+ and π^- almost always is closely correlated with the position of the minimum of the differential cross section. Typically it occurs somewhat before the cross section is at its first minimum.

An examination of the relevant Legendre coefficients in the expansions for I_0 and $I_0 P$, i.e.,

$$I_0 = \hat{\lambda}^2 \sum_{l=0}^{2l_{\max}} A_l P_l(\cos \theta)$$

$$I_0 P = \hat{\lambda}^2 \sum_{l=0}^{2l_{\max}-1} B P_l^{\,1}(\cos \theta)$$

is often useful in trying to get a rough idea of some of the main qualitative features of the data. The quantum number assignments for the $\Delta(1924)$ as F_{37}[12] and the $N^*(2190)$ as G_{17}[13] were initially based on a detailed study of these coefficients. Generally speaking, however, the very large number of partial waves involved in these scattering processes make it difficult, if not impossible, to untangle all of the various effects from a study of the Legendre coefficients alone. It is really only through detailed phase shift analyses of the data that the behavior of the partial-wave amplitudes with energy can be established.

D. Experimental Outlook

Although this is primarily a theoretically oriented conference, a few words are in order about experimental developments. I will be brief. It is well known by now that the use of polarized targets has been responsible for much of the experimental progress in πN scattering. Two comments are in order. First, targets containing significantly more free protons are being developed and should supplant the so-called LMN targets in many of these experiments in the near future. The new targets contain about five times as much free hydrogen as the 3% contained in the LMN crystals. For example, ethyl alcohol seems to be a very suitable material. Second, there are several groups trying to develop hydrogen-rich polarized targets[14] using the so-called "brute force" technique rather than the dynamic method used in existing targets. Here the object is to make the Boltzmann factor, $\mu_p H/kT$, as large as possible by using very high fields ($B \simeq 105$ G), and low temperatures ($T < 10^{-2}$ °K). These developments are a little further off in time, but if realized should find interesting applicability also in πN scattering.

As mentioned previously, a target has been constructed[15] in which the protons can be polarized in the plane of the scattering. Consequently A_{recoil} and R_{recoil} results should be forthcoming in the foreseeable future.

One other technique deserves a brief comment: wire spark chambers. With these detectors very precise position measurements are possible. The state of the technology here is advancing fast, and it seems likely that the very precise measurements of angles which can be made with these chambers will permit very detailed measurements of angular distributions. Automatic read out systems are available which should make it possible to do high statistics experiments of very high quality for both elastic and inelastic scattering.

E. Other Experiments

Mainly because of time limitations I will not discuss here the very interesting and extensive results of the photoproduction experiments. It should be kept in mind, however, that these data are very relevant to any discussion of baryon resonances. There has perhaps not been enough effort devoted to making full use of *all* available data in trying to extract meaningful results from existing experiments. We have become too accustomed to analyzing the elastic scattering experiments quite independently of the photoproduction results and vice versa.

The experimental situation with respect to inelastic πp scattering in this energy region will be discussed by Rosenfeld. (See the paper by Rosenfeld and Söding, this volume.) Detailed information, especially at the lower energies, will provide useful bounds on some of the existing phase shift solutions. Partial-wave analyses of these experiments have the drawback that they are all at least to some degree model dependent. Nevertheless these analyses can be used to check the existing phase shift solutions and vice versa. I will defer the discussion of this subject to Rosenfeld.

III. Phase Shift Analyses

A. Formalism and Notation

I had hoped not to have to go through this again, but there are still some heretics around who refuse to accede to the dogma of the Basel Convention. They—and THEY know who THEY are—insist on defining $\hat{n}$ as $\hat{k}_f \times \hat{k}_i$. Burning at the stake is far too mild a punishment for such a heinous offense. THEY should be buried alive in ream after ream of the computer printout of their phase shift analysis. Then there are others who despite howls of anguish from their tortured brethren insist on expressing their phase shift results as a function of pion kinetic energy. It is difficult to conceive of a severe enough penalty for the perpetrators of such a sin. Perhaps we should bug their CDC 6600 so that it will ignore the middle 5 bits of all their computer words when they make their analysis. On second thought, this may not be such a good idea after all—they might get even better fits to their dispersion relations. At any rate let me make a plea for some standardization: Let us use total energy in the center of mass to specify the phases and the resonant states, and whenever possible also indicate the pion momentum in the lab. With respect to $\hat{n}$, can we all agree to use $\hat{n} = \hat{k}_i + \hat{k}_f$ where $\hat{k}_i$ and $\hat{k}_f$ refer to the direction of the meson in the center of mass before and after the scattering?

The partial-wave decomposition of the scattering amplitudes can be written

$$G = \frac{1}{k} \sum_{l=0}^{\infty} [(l + 1)T_{l+} + lT_{l-}]P_l(\cos \theta)$$

$$H = \frac{1}{k} \sum_{l=1}^{\infty} [T_{l+} - T_{l-}]P_l^{1}(\cos \theta)$$

where $l\pm$ stands for $j = l \pm \frac{1}{2}$ and $T_{l\pm} = [\eta_{l\pm}\exp(2i\delta_{l\pm}) - 1]/2i$ are the partial wave amplitudes; $\eta_{l\pm}$ is the absorption parameter ($\eta_l = 1$ corresponds to no absorption, $\eta_l = 0$ corresponds to complete absorption); and $\delta_{l\pm}$ is the phase shift for the state $j = l \pm \frac{1}{2}$. The partial wave amplitudes $T_{l\pm}$ are conveniently represented in graphical form by an Argand diagram (see Fig. 8).

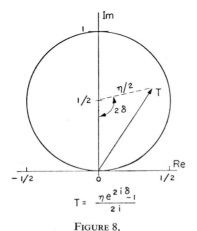

FIGURE 8.

A simple Breit-Wigner resonance can be written

$$T_e = x/(\varepsilon - i)$$

with $x = \Gamma_{\text{elastic}}/\Gamma_{\text{total}}$ (the elasticity of the resonance which is not the same as absorption parameter η) and $E_{\text{Res}} =$ the resonance energy; $E =$ the energy; $\Gamma =$ the total width of the resonance. In this representation such a resonant amplitude would describe a circle moving counterclockwise as the energy E increases. When $E = E_{\text{Res}}$ the resonant amplitude is pure imaginary. Thus, if there is no background, a resonant amplitude will have $\delta = 0°$ or $90°$ depending on whether $x < \frac{1}{2}$ or $x > \frac{1}{2}$ (see Fig. 9). Often a resonance amplitude is superposed on some background. In that case the resonant circle will not originate at the origin but somewhere else within the unitary circle.[16] Other factors such as energy-dependent widths will further distort the picture of a smooth circle.

The problems of the phase shifter are thus twofold. In the first place he must try to find a unique set of phase shifts which satisfactorily fit all the data at all energies. Then he must endeavor to extract the various resonance parameters from these results. For those partial wave amplitudes which describe a reasonable facsimile of a circle when plotted in an Argand diagram, this problem is not too bad; but when the partial-wave amplitudes wobble and stagger their way through the unitary circle, it is often very difficult to determine the resonance parameters, and sometimes even to know if the amplitude resonates or not.

In presenting phase shift results we will use the usual notation for the partial-wave amplitudes and phases: $L_{2T,2J}$. For example S_{31} refers to $l = 0, T = \frac{3}{2}, J = \frac{1}{2}$.

B. Methods

Most of the groups involved in the phase shift business use slightly different methods to obtain their results. These differences concern mainly the extent to which assumptions about the variation of phase shifts with energy are put into the analysis *a priori*. The various methods can be divided into two broad classes—energy dependent and energy independent.

In the energy dependent approach one has some preconceived ideas of what nature must be like. One puts these ideas into what is usually a some-what oversimplified mathematical form involving lots of free parameters including the phase shifts. One analyzes all data at all energies simultaneously with the help of a computer to find the set of these parameters which best fit the data. If the fit is satisfactory one has a solution—if not one usually throws out the parametrization used (or sometimes even the theory) and tries something new and so on. The main proponents of this method are the Livermore[17] and Chilton (now Durham) groups.[18]

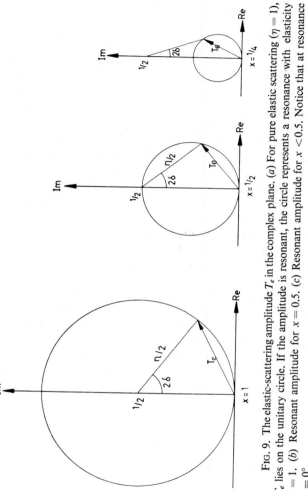

FIG. 9. The elastic-scattering amplitude T_e in the complex plane. (a) For pure elastic scattering ($\eta = 1$), T_e lies on the unitary circle. If the amplitude is resonant, the circle represents a resonance with elasticity $x = 1$. (b) Resonant amplitude for $x = 0.5$. (c) Resonant amplitude for $x < 0.5$. Notice that at resonance $\delta = 0°$.

13

In the energy independent phase shift searches only the experiments within a very narrow interval of energies are analyzed at any one time. The number of possible acceptable solutions at each energy depends on the quality and quantity of the experimental data as well as the number of partial waves used in the analysis. In general there are many solutions at the higher energies. For example, at $P_\pi = 1.0$ GeV/c the Berkeley group has at least 45 solutions which can be considered statistically acceptable fits to the data. At energies below about 400 MeV the solution obtained at each energy is thought to be unique. In this approach the sticky problem comes when one tries to make a continuation from one energy to the next. In trying to choose the "correct" solution at each energy each of the groups uses some sort of smoothness criterion to constrain the variation of the phase shifts with energy; i.e., the phase shifts or the amplitudes or the observables are expected to vary rather smoothly with energy—no *sharp* discontinuities are expected in any of these quantities. The CERN group[20] makes the energy continuation with the help of dispersion relations for the partial-wave amplitudes. They go so far, in fact, that at a certain stage in their analysis the dispersion relation constraints become part of the "input data" (along with the experimental points) which the phase shifts must fit. Consequently the CERN solution is very much smoother than any of the others; their dispersion relations force their solutions to behave smoothly with energy. I am sure Dr. Lovelace will elaborate on the methods used by the CERN group in his paper. The Saclay[19] and Berkeley[21] groups are less sophisticated (a carefully chosen word, in this case meaning that in reality we are probably still too stupid to fully appreciate the power of dispersion relations in this type of work), and consequently we use our eyeballs or the computer to select the "smoothest" combination of solutions. This latter approach has the advantage that solutions tend to be less biased by theory. As Lovelace has pointed out, it is much better to be biased by a "good" theory than to be unbiased; nevertheless, the question which always lurks in the background concerns how valid the theory really is. Keep in mind that there is a bias in all of these approaches—the assumption of "smoothness" biases us against finding very narrow resonances or cusp-like phenomena. By hard experience we have learned that it is very difficult indeed to find a unique solution without some theoretical input. The Saclay and Berkeley solutions tend to be much less smooth than that of the CERN group. There is a lot of fine structure which is probably not significant but which neither the Saclay nor the Berkeley groups have been able to eliminate in their energy-independent, no-theory-input approach. It seems to me that the input of some theoretical assumptions cannot be avoided completely if one wants to get a "reasonable" solution to this problem. I think the approach of the CERN group is probably the closest approximation so far made to what should be done, although I would favor less reliance on the partial-wave dispersion relations for the low partial-wave amplitudes than they used.

There is one other energy-independent analysis which should be mentioned. It is the one of the Hawaii group.[22] Here an attempt was made to see if the existing data up to 700 MeV ($P_\pi = 830$ MeV/c) could be fit with amplitudes which do not resonate. It was a sort of euthanasia campaign launched by Cence to see if he could curb the population explosion of πN resonances. It was an interesting exercise which indicated that a reasonably good fit could be obtained to all of the data with a set of amplitude shifts which do not resonate (with the exception of $\Delta(1236)$). Closer examination of these results showed, however, that they were in contradiction with dispersion relations for the spin-flip amplitudes, and I believe that the more complete experimental information now available cannot be satisfactorily fit by this solution.

In Table I, I have summarized the main features of the methods used and the results obtained in the various analyses which have so far been reported. I will leave it to Dr. Lovelace to give us his usual tempered assessment of the relative merits of each of these efforts.

D. New Results

In recent months three groups—Saclay, CERN, and Berkeley—have reported the latest versions of their phase shift analyses. The relevant Argand diagrams for each of these solutions as well as a table summarizing the present status of πN resonance parameters is shown in Table II and Figs. 10 and 11. For the true aficionados we have even prepared a limited number of copies of the detailed results of each of these groups. These are available from the authors while the supply lasts. Time does not permit me to discuss each of these analyses in detail. What I would like to do, instead, is to point out some of the main characteristics of each, and then to discuss some of their similarities and differences.

Let's begin with Saclay. Their new results differ from their earlier work[23] in that they are based on more reliable experimental data and new polarization measurements. The new analysis goes up to 2025 MeV total mass ($P_\pi = 1700$ MeV/c). The maximum orbital angular momentum used in $l_{max} = 4$ (G waves). They obtained $\chi^2 = 1633$ for 1433 degrees of freedom.[24] The variation of the phase shifts with energy is quite bumpy for many of the partial waves, i.e., there seems to be some fine structure which they are unable to get rid of. It is not clear at present what, if any, significance should be attached to this fine structure. What is clear, is that up to now they have not been able to get rid of it even though they have tried pretty hard. Their results, which have been published in the *Physical*

TABLE I

Summary of Main Features of Various Phase Shift Analyses

Group	Method	Input assumptions	Energy range	Comments
Livermore[a]	Energy dependent	$\delta_l = k^{2l+1} \sum\limits_{n=0}^{n_{max}} A_n k^n$ + Breit-Wigner resonances	Up to $T_\pi = 800$ MeV ($P_\pi = 930$ MeV/c, $M^* = 1630$ MeV)	First found P_{11} (1470) "Roper" resonance. Assumptions about energy dependence of phases tends to bias against finding new resonances. Results published.
Saclay[b]	Energy independent	None	Up to $T_\pi = 1.6$ GeV ($P_\pi = 1.7$ GeV/c, $M^* = 2025$ MeV)	Many solutions at each energy. Energy continuation made by making smooth connection between phase shifts at different energies. First found complex resonant structure in regions of (1512) and (1688) resonances. Work completed.
Chilton[c]	Energy dependent	Energy continuation made with help of dispersion relations for inverse amplitudes	Up to $T_\pi = 1.1$ GeV ($P_\pi = 1.23$ GeV/c, $M^* = 1795$ MeV)	Some results published. Work in progress.
Hawaii[d]	Energy independent	Assumed no resonances	Up to $T_\pi = 700$ MeV ($P_\pi = 830$ MeV/c, $M^* = 1575$ MeV)	Wanted to see if existing data could be satisfactorily fit with non-resonant amplitudes. Found solution which is in reasonable agree-

				ment with experimental observations. However, results disagree with spin-flip dispersion relations and with newer, more extensive data.
CERN[e]	Essentially energy independent analysis but with dispersion relation input	Energy continuation made with help of dispersion relations for partial-wave amplitudes	Up to T_π = 1.94 GeV (P_π = 2.07 GeV/c, M^* = 2190 MeV)	Most sophisticated analysis. They check self-consistency of dispersion relation input. Results could be slightly biased because solutions are forced to be in accord with dispersion relation input. Have found 18 resonant states. Results published.
Berkeley[f]	Energy independent	None	Up to T_π = 1.6 GeV (P_π = 1.7 GeV/c, M^* = 2025 MeV)	Many solutions at each energy. Energy continuation based on smooth variation of amplitudes made with help of computer. Uniqueness of solutions not established. Work in progress.

[a] L. D. Roper, R. M. Wright, and B. T. Feld, *Phys. Rev.*, **138**, B190 (1965).

[b] P. Bareyre, C. Bricman, and G. Villet, *Phys. Rev.*, **165**, 1730 (1968).

[c] B. H. Bransden, P. J. O'Donnell, and R. G. Moorhouse, *Proc. Roy. Soc.* (*London*) *Ser. A*, **289**, 538 (1966). (300–700 MeV); *Phys. Rev. Letters*, **19**, 420 (1965) (700–1100 MeV).

[d] R. J. Cence, *Phys. Rev. Letters*, **20**, 306 (1966).

[e] A. Donnachie, R. G. Kirsopp, and C. Lovelace, CERN Internal Report TH.838 (1967).

[f] C. H. Johnson, thesis, University of California, Berkeley, UCRL–17683 (1967).

TABLE II

π-N Resonance Parameters (in MeV)

Wave[a]	Mass	Γ_{tot}	Γ_{el}/Γ_{tot}
P_{33}	1235.8	125.1	1.0
P_{11}	1470	210	0.65
D_{13}	$\sim$1520	115	0.55
S_{11}	1535	120	0.35
S_{31}	1640?	180	0.30
D_{15}	1680	170	0.40
F_{15}	1690	130	0.65
$*P_{33}$	1688	280	0.10
$*D_{33}$	1691	270	0.14
S_{11}	1710	300	0.80
$*P_{11}$	$\sim$1751	330	0.32
$*P_{13}$	$\sim$1863	$\sim$300	$\sim$0.21
$*F_{35}$	1913	350	0.16
$*P_{31}$	1934	340	0.30
F_{37}	1950	220	0.40
$*D_{35}$	$\sim$1954	$\sim$310	$\sim$0.15
$*F_{17}$	1983	225	0.13
$*D_{13}$	2057	290	0.26
G_{17}	$\sim$2200	300	0.35

[a] Asterisks mean new resonance. The parameters of the new resonances are taken from A Donnachie, R. G. Kirsopp and C. Lovelace, CERN TH. 838. Those of the more established resonances are obtained from weighing the results of the Berkeley, Saclay, and CERN groups.

Review,[19] represent the final version of this stage of the analysis. No further work is contemplated in the immediate future.

The Berkeley results are not finished. They represent the present status of an analysis which is in progress. The experimental data used are substantially the same as those used by the Saclay group. Although $l_{max} = 4$ also in the case of this work the actual number of parameters used to fit the data was slightly larger because five additional parameters were used to normalize the five different types of experiments at different energies. In this way systematic errors resulting from improper normalization of the reported experimental measurements could be taken into account. At present this analysis, like that of Saclay, extends to a total mass

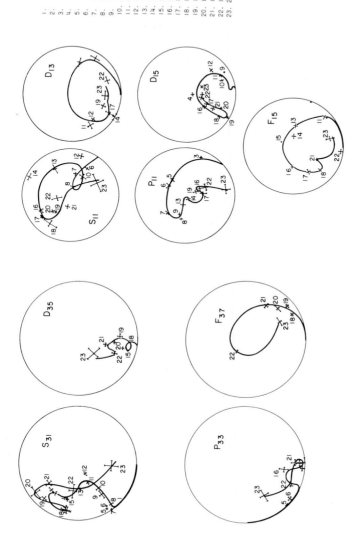

E_{cm}	
1.	1320
2.	1362
3.	1390
4.	1443
5.	1470
6.	1501
7.	1524
8.	1543
9.	1573
10.	1603
11.	1617
12.	1629
13.	1658
14.	1673
15.	1688
16.	1716
17.	1738
18.	1769
19.	1822
20.	1862
21.	1896
22.	1968
23.	2021

FIGURE 10. Partial wave amplitudes obtained by the saclay phase shift analysis (Bareyre et al.).

19

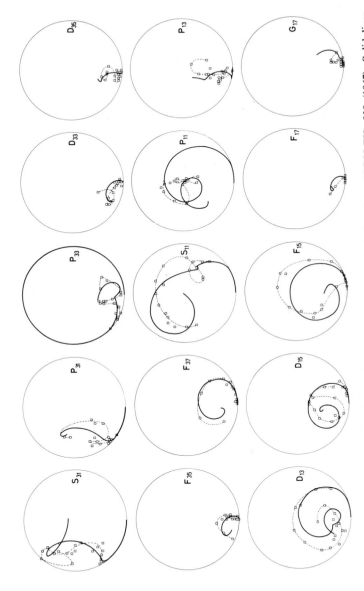

FIGURE 11. Taken from A. Donnachie, R. G. Kirsopp, and C. Lovelace, CERN TH. 838 (1967). Solid line: Partial wave amplitudes obtained from the dispersion relation results of the CERN group. Dashed line: Partial wave amplitude obtained from the Berkeley group.

of 2025 MeV ($P_\pi = 1700$ MeV/c). Preliminary results have been reported.[21] $\chi^2 = 1790$ for 1466 degrees of freedom.[24] The variation of the phase shifts with energy is comparable in "bumpiness" with that of Saclay, although there are noticeable differences in the behavior of several of the partial-wave amplitudes between the Berkeley and Saclay results. Although the procedure used to determine the continuation of the partial-wave amplitudes with energy was quite different in the case of Berkeley and Saclay (i.e., Berkeley used the computer to impose a certain smoothness condition while Saclay imposed quite a different condition and did it more or less by hand) most of the results are very similar.

The CERN analysis is certainly the most impressive of the lot. Their results extend to a mass of 2190 MeV ($P_\pi = 2070$ MeV/c). They go to $l_{max} = 4$ except at their highest energies where H waves are used. Some care should be exercised in using these results. There are two sets of phase shifts quoted. One is the "experimental" set, which is based on the fits to the experimental data plus the additional constraints imposed by the dispersion relations. The second is the "dispersion relation" set, which is simply the set calculated directly from the dispersion relations themselves. This latter set by its very nature is exactly as smooth as the dispersion relations specify. The "experimental" phases are somewhat bumpier, but even they seem to be strongly enough constrained by the dispersion relations in most cases so as to be much, much smoother than anything Saclay or Berkeley has to offer. The quoted χ^2 is 4421 for 4102 degrees of freedom.[24] The disparity in the χ^2 values and the number of degrees of freedom[24] between the various results comes mainly from the fact that the CERN group includes much more low energy data in their analysis than Berkeley and Saclay. The CERN results are published in preprint form[20] and have been reported at the Heidelberg Conference.

As a result of their phase shift analysis the CERN group has proposed a total of 18 resonances including 9 previously undiscovered ones in the mass region below about 2300 MeV. The resonances in Table II which are not preceded by an asterisk are considered to be well established. Some of these, like the $P_{33}(1238)$, $D_{13}(1520)$, $D_{15}(1680)$, $F_{15}(1690)$, $F_{37}(1950)$, $G_{17}(2200)$, $S_{11}(1710)$, $P_{11}(1470)$ are beyond all doubt, although the exact resonance parameters may well change slightly. Where the energy regions of all three analyses overlap these resonances are present in all of them. Sometimes the exact behavior of the amplitudes in the Argand plot is somewhat different, as for example in the case of F_{37} where CERN and Berkeley are very similar, but Saclay is noticeably different. There is also little doubt that S_{31} resonates at about 1640 MeV, although the behavior of this amplitude is rather strange. The first S_{11} resonance seems to be present in the Saclay and Berkeley results, but missing in the case of CERN.

Interestingly enough, when the CERN results are used as input to an unconstrained analysis (i.e., no dispersion relations) in which the Berkeley data set is fitted, the S_{11} resonance near 1540 MeV reappears. Although this may not be a completely fair way to play the game it is worth pointing out that the very smooth CERN phases all readjust themselves, and in so doing they become just as bumpy as the Berkeley and Saclay results. It seems that the dispersion relations are powerful medicine indeed against such hiccups.

The resonances which are preceded by an asterisk are the new resonances uncovered by the CERN group.[20] It is premature to consider all of them as well established. Certainly the general behavior of several of these amplitudes such as, for example, F_{35}, D_{33}, P_{31}, and $P_{33}(1688)$ have common resonance-like features in the CERN and Berkeley analyses. I believe that the chances are good that these resonances are real. On the other hand, it takes a lot of imagination and probably some wishful thinking to see resonance-like behavior in the Argand plots of D_{35} and P_{13}. Even if all of these resonances were to survive it seems likely that the quoted resonance parameters may well undergo significant changes as more information becomes available. *These remarks are intended as a word of caution to those who would take these results too seriously.* One of the reasons that it is difficult to draw firm conclusions about many of these new resonances is that they tend to be quite inelastic; i.e., their coupling to the elastic channel is small. Consequently they do not manifest themselves as obviously as one might like in an analysis of elastic scattering experiments. The detailed inelastic scattering experiments now in progress should be of great help in resolving these questions.

A question which often comes up in a discussion like this is: "Are these phase shift solutions really unique, or are there other equally good solutions lying undiscovered somewhere in the bowels of the 6600?" The CERN and Saclay groups believe that their solutions *are* unique. That does not mean that the existing solutions cannot undergo some readjustments. It does mean, however, that these groups expect the general features of their results not to change significantly. The Berkeley group is much less sure that there are not other solutions in the energy region spanned by their analysis. For example, only a few days before this conference, they obtained another solution up to ∼1120 MeV/c (center-of-mass energy = 1740 MeV) which fits the data even slightly better than the solution quoted here and whose behavior differs in several respects from the usual solution. These differences concern mainly the elasticities of the $T = \frac{1}{2}$ resonances and the behavior of P_{11} above the 1470 MeV resonance. It is too early to tell if this new solution will ultimately also fit the data at higher energies. It is possible that it is only a variant of the existing solution. So far as the Berkeley group is concerned the question of uniqueness is still open.

In discussing the question of uniqueness one must specify some conditions which determine when two solutions are really different and when they are the same. Generally speaking, the errors associated with the phase shifts are smaller than most of us feel they have any right to be. That is, there are often solutions which differ in several of the phases by many standard deviations—and still we think the solutions are essentially the same. It is often as though the hypersurface near the χ^2 minimum in the phase shift space is covered with little " pot holes." Perhaps the assumption of hyperparabolic behavior of χ^2 as a function of the phase shifts near χ^2_{min} is questionable in such a case. At any rate, two solutions are usually thought to be the same unless the general behavior of at least one of the phases is *markedly* different from one solution to the other. *Again here I would caution the uninitiated to consider the quoted phase shift errors to be smaller in general than the actual uncertainty associated with our knowledge of these parameters.* As an example of this fact, consider the " fine structure " observed in the Saclay and Berkeley analyses. This fine structure is usually much larger than the errors, and yet we still don't know if such fine structure is *really* significant. Perhaps we just have to try a little harder if we want it to go away.

Let me conclude this survey with the most spectacular result of all. Most of you are familiar with the view espoused in certain quarters on the

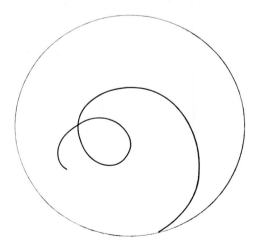

Fig. 12. Argand diagram for partial wave amplitude P_{11} obtained by the CERN group.[20]

fifth floor of Building 50B in Berkeley, that the bubble chamber represents the ultimate tool of the research physicist. It took a while, but I have finally become a believer. And many of you will, too, once you have seen what I am about to show you. Recall the beautiful results of the CERN group such as those shown in the Argand diagram for P_{11} (Fig. 12). With due apologies to Dr. Lovelace, I must state that his results pale into insignificance compared to the result I am about to report. It is an Argand plot of P_{11} from the threshold to *infinity* taken in a bubble chamber. (You can just hear the mouths of the Reggeists water at the thought.) The nine new P_{11} resonances are shown in Fig. 13.

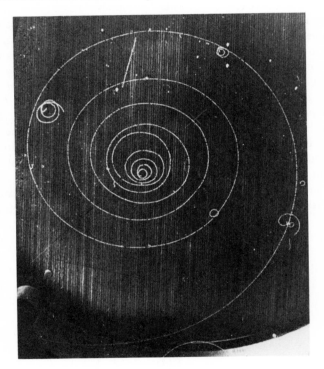

FIGURE 13.

References

1. L. Wolfenstein and J. Ashkin, *Phys. Rev.*, **85**, 947 (1952).
2. W. Busza, D. G. Davis, B. G. Duff, F. F. Heymann, C. C. Nimmon, D. T. Walton, E. H. Bellamy, T. F. Buckley, P. V. March, A. Stefanini, J. A. Strong, and R. N. F. Walker, *Nuovo Cimento*, **52**, 7823 (1967).

3. University College, Westfield College London Collaboration, Contributed paper submitted to the Heidelberg International Conference on Elementary Particles, 1967.

4. A. S. Carroll, J. Fischer, R. H. Phillips, C. L. Wang, A. Lundy, F. Lobrowicz, A. C. Melissinos, Y. Nagashima, and S. Tewksbury, $\pi^{\pm}p$ Backward Scattering Between 1.5 and 3.0 BeV/c, contributed paper submitted to this conference.

5. Cambridge, Rutherford, Birmingham Collaboration, Pion–Nucleon Total Cross Sections From 0.5 to 2.65 GeV/c, Rutherford Laboratory Preprint RPP/H/32.

6. C. R. Cox, K. S. Heard, J. C. Sleeman, P. J. Duke, R. E. Hill, W. R. Holley, D. P. Jones, F. C. Shoemaker, J. J. Thresher, and J. B. Warren, Rutherford Report No. RHEL/M 137.

7. W. Gorn, C. C. Morehouse, T. Powell, P. R. Robrish, S. Rock, S. Shannon, H. Steiner, and H. Weisberg, *Bull. Am. Phys. Soc.*, **12**, 469 (1967), and W. Gorn, private communication.

8. R. E. Hill, N. E. Booth, R. J. Esterling, D. L. Jenkins, N. H. Lipman, H. R. Rugge, and O. T. Vik, *Bull. Am. Phys. Soc.*, **9**, 410 (1964).

9. D. D. Drobnis, J. Lales, R. C. Lamb, R. A. Lundy, A. Moretti, R. C. Niemann, T. B. Novey, J. Simanton, A. Yokosawa, and D. D. Yovanovitch, Measurement of Polarization in $\pi^- p \rightarrow \pi^0 n$ and $\pi^- p \rightarrow \eta n$, reported at this conference.

10. O. Chamberlain, M. J. Hansroul, C. H. Johnson, P. D. Grannis, L. E. Holloway, L. Valentin, P. R. Robrish, and H. M. Steiner, *Phys. Rev. Letters*, **17**, 975 (1966); M. Hansroul, Measurement of the Polarization Parameter in the Elastic Scattering of Negative Pions by Protons From 600 to 3300 MeV/c, (Ph.D. Thesis), Lawrence Radiation Laboratory Report UCRL-17263, January 1967.

11. P. Bareyre, C. Bricman, and G. Villet, Saclay report No. CEA-R3401.

12. P. J. Duke, D. P. Jones, M. A. R. Kemp, P. G. Murphy, J. D. Prentice, J. J. Thresher, H. H. Atkinson, C. R. Cox, and K. S. Heard, *Phys. Rev. Letters*, **15**, 468 (1965).

13. A. Yokosawa, S. Suwa, R. E. Hill, R. J. Esterling, and N. E. Booth, *Phys. Rev. Letters*, **16**, 714 (1966).

14. A. Honig, *Phys. Rev. Letters*, **19**, 1009 (1967). E. Varoguaux, ^{3}He–^{4}He Dilution Refrigerators, *Proc. Intern. Conf. Polarized Targets Ion Sources, Saclay, 1966*; B. S. Neganov, private communication, 1967.

15. L. Van Rossum, private communication, 1967.

16. R. H. Dalitz, *Ann. Rev. Nucl. Sci.*, **13**, 339 (1963); C. Michael, *Phys. Rev. Letters*, **21**, 93 (1966).

17. L. D. Roper, *Phys. Rev. Letters*, **12**, 340 (1964); L. D. Roper and R. M. Wright, *Phys. Rev.*, **138**, B921 (1965).

18. B. H. Bransden, R. G. Moorhouse, and P. J. O'Donnell, *Phys. Rev.* **139**, B1566 (1965); B. H. Bransden, R. G. Moorhouse, and P. J. O'Donnell, *Phys. Letters*, **11**, 339 (1964); B. H. Bransden and P. J. O'Donnell, *Phys. Letters*, **19**, 420 (1965).

19. P. Bareyre, C. Bricman, and G. Villet, *Phys. Rev.*, **165**, 1730 (1968).

20. C. Lovelace, Nucleon Resonances and Low Energy Scattering, Rapporteur's talk at the 1967 Heidelberg Conference, CERN Report No. TH-837, October, 1967; A. Donnachie, R. G. Kirsopp, and C. Lovelace, Evidence from πp Phase Shift Analysis for Nine More Possible Nucleon Resonances, CERN Report No. TH-838, October, 1967.

21. C. H. Johnson, Jr., Measurement of the Polarization Parameter in $\pi^+ p$ Scattering from 750 to 3750 MeV/c, (Ph.D. Thesis) Lawrence Radiation Laboratory Report UCRL-17683, August, 1967.

22. R. Cence, *Phys. Letters*, **20**, 306 (1966).

23. P. Bareyre, C. Bricman, A. V. Stirling, and G. Villet, *Phys. Letters*, **18**, 342 (1965).
24. In defining the total number of degrees of freedom here we have simply added together the number of degrees of freedom (i.e., number of measurements minus number of phase shift parameters) at each of the energies used in the analysis. This number is certainly an underestimate for the total number of degrees of freedom because the phase shift parameters at different energies are correlated. Consequently conclusions about the goodness of these fits based on statistical considerations alone tend to be misleading. It is very difficult to make a precise statement about the statistical significance of the various analyses described here.

πN Phase Shift Analysis and Phenomenological Dispersion Relations

C. LOVELACE

CERN, Geneva

1. Some Remarks on Resonances

I have left it to Herb Steiner to comment[1] on the nine new resonances we announced recently,[2] since he needed some light relief from the guts of polarized targets. I had my say at Heidelberg[3] and didn't intend to repeat myself here. However, since then the amount of πp data in this energy region has shot up by about 50%, due especially to the vast new Chilton $\pi^- p$ polarization experiment[4] you heard about this morning. Thresher has a graph showing his data against the CERN[2] and Saclay[5] phases, neither of which included it. I don't think I am being biased when I say that it appears to me to favor the former. The disagreement with the Saclay phases arises mainly from near the forward direction (where there were no previous measurements). It is probably, therefore, the high partial waves in the two solutions which are being tested.

We have now included all these new data in our fit, together with the Carroll et al. backward cross sections,[6] the Chilton total cross sections,[7] and others. All our nine new resonances are still there! In fact, except around 1500 MeV mass, the phases have hardly changed. The main difference there is that we now have the lower S_{11} resonance like other people, and that $D_{13}(i)$ has moved back to 1520 MeV.

My own view is that, except for S_{11} and perhaps P_{11}, the phase shifts are now fairly definitive up to 2000 MeV mass. I am less happy about the 2000–2190 MeV region, where there are still serious gaps in the experiments. This is why the number of new resonances we claimed was only 9 and not 14, which it might well have been in the hands of a less conservative person than myself, as can be seen from a recent paper of Barger and Cline.[8]

On resonances that are not seen, the evidence against $F_{37}(1946)$ having a nearby G_{37} partner with any appreciable coupling to the elastic channel seems to me very strong. Close parity doubling on the leading Regge trajectory will be very easy to detect. In fact, it will show up in polynomial fits without even doing phase shift analysis.

27

Most of the new resonances are seen against large backgrounds, which makes the mass determination rather delicate. Fortunately, the background usually seems to be predominantly elastic. Under these circumstances we can decompose phase and elasticity[9]

$$\delta(W) = \delta_R(W) + \delta_B(W) \tag{1}$$
$$\eta(W) = \eta_R(W) \times \eta_B(W) \tag{2}$$

and $\eta_B(W)$ will be slowly varying and close to 1. $\delta_B(W)$, on the other hand, may change rapidly. This makes it advisable to determine the resonance masses and widths from η alone. The mass m will then be the point at which η achieves its minimum η_m, and the half-width $m - \Gamma/2$ will be the point where

$$\eta[m - (\Gamma/2)] = \sqrt{[(\eta_m^2 + \eta_B^2)/2]^{1/2}} \tag{3}$$

The elastic branching ratio will be

$$\Gamma_{el}/\Gamma_{tot} = \tfrac{1}{2}(\eta_B - \eta_m) \tag{4}$$

The numbers in the written version of my Heidelberg talk were obtained from these formulas.

The phase shift analyses are based on formation experiments. Production experiments often see nucleon resonances with statistics that were surpassed in the formation experiments ten years ago, but this does not seem to stop their authors from claiming tiny errors on their resonance parameters, based on fits with a pure Breit-Wigner over a phase space background, and then supposing that these refute the phase shift analyses. Their errors are small because their statistics are small enough to fit with naive models. I think that, in the present state of the art, anyone who quotes a statistical error on the mass of a nucleon resonance is a liar.

2. Phase Shift Analysis

a. The Problem

Johnson and Steiner[1] have already presented notations and formulas. Phase shift analysis consists then in solving these complicated nonlinear equations that express the observables in terms of the phases and elasticities of the different partial waves. Because of their nonlinearity and the incompleteness of the experiments there will usually be many solutions at each energy. However, it has been found in practice that most of this ambiguity can be eliminated by doing phase shift analysis at a number of energies and using continuity between the different energies to select among the solutions found. Phase shift analyses go out of date after a year or two because of improved experiments. All the three current ones (Saclay,[5] Berkeley,[10] and CERN[2]) were based on this continuity technique. At the highest energies thus far studied, S, P, D, F, G, and H waves are all needed, giving 44 variables. In principle, it would be better to do energy-dependent fits to the data from all energies simultaneously, but more variables would then naturally be required. The largest existing energy-dependent fits actually have fewer variables than the largest single-energy fit, so it is not surprising that they were statistically unsatisfactory. However, I expect 200 variable fits to become a practical proposition within the next year or two (more as a result of programming improvements than of bigger computers), and energy-dependent fits may then be restored to life. Even with such superprograms, it appears likely that single-energy solutions will be needed as the first stage of the search.

b. Finding the Solutions

The only way to find all the solutions at a particular energy is by random searches. This means that the phases are started from some random set of values, and χ^2 is then minimized by standard numerical methods.[11] This process is then repeated from hundreds of different random starting points until each minimum has been found so often that it seems likely there are no more. Success here depends primarily on the speed of the computer program. When each minimum is reached in seconds, much more thorough searches are possible than if it takes hours. Some programs are 1000 or more times faster than others, and aspiring phase shift analysts had better learn about such matters if they want to stay alive.

Phase shift analysis is only worthwhile because the centrifugal barrier effectively cuts off the partial-wave expansion. The number of partial waves required increases with energy, and so therefore does the quantity of measurements required for a phase shift analysis. If there are too few experiments or if too many partial waves are included, then the number of solutions becomes infinite. Phase shift analysis is therefore impossible without either assuming higher partial waves to be zero or else taking them from dispersion relations. It is not apparent to me that the former procedure is in any way less biased than the latter. Moreover, it can be shown both numerically and analytically that small changes in the high partial waves will induce big changes in certain combinations of the low ones. For example, in $\pi^+ p$ scattering below 200 MeV, only S and P waves can be determined from the experiments, yet it can be shown analytically that S_{31} is unstable against small changes in the D waves. The only safe course is then to determine the D waves from phase shift analysis above

200 MeV and extrapolate down. Therefore no energy region can be securely pacified until outposts have been established in the region above. The top energy end of any phase shift analysis will be relatively uncertain. This is just the old imperialist situation: we are continually forced to expand our territory in order to protect our boundaries.

c. Linking the Solutions

At the higher energies, even the ones with the best experiments often have 30 or more solutions each. Each solution has 36 parameters in an *SPDFG* fit. Selecting the solutions at the different energies in such a way as to make all these parameters vary continuously, is therefore highly non-trivial. I shall only describe the Berkeley technique[12] for doing this, since it is clearly superior to the others.

First, one needs a measure of the distance between two solutions. They take

$$\Delta(1, 2) = \left\{ \sum_{l,l\pm} |T_{l\pm}{}^{I}(1) - T_{l\pm}{}^{I}(2)|^2 \right\}^{1/2} \tag{5}$$

i.e., the distances between the real parts and between the imaginary parts for each wave are used to define a Euclidean space. Probably these distances ought to be weighted because the statistical errors are smaller on the higher partial waves, but this is a minor point. Now they assume that at some low energy a unique solution is known, and they calculate the shortest path passing through one solution at each intermediate energy and ending in a particular solution at the highest energy. These can be calculated iteratively, since the shortest path to any solution at the kth energy must be a continuation of the shortest path to *some* solution at the $(k - 1)$th energy. Therefore, if there are K energies and N solutions at each, the total number of paths is N^K, but the number that need to be examined to find the shortest path to each solution at the top energy is only $(K - 1)N^2$. Since in the Berkeley analysis $K = 23$ and $N \sim 30$, this is a rather considerable reduction.

The Berkeley group found that the shortest paths to any solution at the highest energy all tended to pass through the same solutions at intermediate energies. In other words, only one qualitatively different solution at each energy seems capable of being linked into a *long* chain, though there are numerous possibilities for forming short chains. This long chain turns out to have all the previously established resonances, despite the fact that the presence of a resonance will tend to lengthen the path. I understand that they now have an alternative chain which continues through all but the last 6 energies, but which differs mainly in S_{11} and P_{11}.

d. Smoothing the Solutions

The phase shifts that emerge from this linking process are still too bumpy to read off resonance parameters, and sometimes even to be sure what resonances are there. They must therefore be smoothed. At first sight this looks trivial—you fit a Breit-Wigner or some other formula to them, just as you would to a histogram, and if the fit is good statistically you feel happy. Unfortunately the errors on the different partial waves are very strongly correlated, which means that satisfactory smoothed fits to the *separate* partial waves, when put together usually give totally unacceptable fits to the original experiments. It is always necessary therefore to redo the phase shift analysis after the smoothing.

Let us consider these correlations. Suppose our solution gives values v_i for the variables, and E_{ij} is the error matrix. The usual statistical errors are

$$\Delta v_i = E_{ii}^{1/2} \qquad (6)$$

The strong correlation means that E_{ij} has large off-diagonal elements. Since the error matrix is symmetric and positive definite, we can decompose it

$$E_{ij} = \sum_n c_i^n c_j^n \qquad (7)$$

where c_i^n are its eigenvectors, each multiplied by the square root of the eigenvalue. Then for a strong correlation the sum in (7) will be completely dominated by a few long eigenvectors. If the solution is changed by

$$v_i \rightarrow v_i + \sum_n \xi_n c_i^n \qquad (8)$$

χ^2 will increase by

$$\chi^2 \rightarrow \chi^2 + \sum_n \xi_n^2 \qquad (9)$$

If one eigenvector dominates, then $\Delta v_i \sim |c_i^1|$. Now, if one changes all N variables by Δv_i, one usually expects to increase χ^2 by $\sim N$. However, if we choose the signs in the direction of the dominant eigenvector c^1, we see from (9) that this will increase it only by 1. Thus, provided we are careful which way we move, we may well be able to change all the variables by many standard deviations before the fit gets appreciably worse. This is why, as is indicated in the article by Johnson and Steiner, the statistical errors give in some respects an underestimate of the true uncertainty. On the other hand, for the small eigenvectors $c_i^n \leq \Delta v_i$, and if we move along them the fit may disappear completely long before the variables have changed by even one standard deviation. Since there are many more short eigenvectors to hit than long ones, this explains why smoothing can be so destructive. The quoted errors on the phase shifts are thus only underestimates if we are trying to prove uniqueness. For getting rid of wobbles they are overestimates, unless we are lucky enough to hit the right combination. I fear therefore that the discontinuities in some people's phase shifts may be even *more* serious than they look from the error bars.

The long eigenvectors are also the directions in which the shape of the χ^2 valley needs to be more carefully observed, since the chances of it ceasing to be quadratic before the fit has disappeared are higher. Calculation of the leading eigenvectors of the error matrix is extremely useful in picturing correlations and I would urge experimentalists to get used to it.

The moral is that phase shifts are not histograms, and only their authors can really understand what the errors on them mean.

In low energy nucleon–nucleon phase shift analysis smoothing has been successfully performed by using the full error matrix on each single energy solution and fitting all partial waves together.[13] However, they had only 6 energies, whereas we have 59 and would need something like 400 variables. A program intended to be capable of such colossal fits is at present under construction at CERN, but our published results were achieved by an indirect method.

We admit, therefore, that our first smoothed fits will be largely wrong. A few waves may be qualitatively correct, but we don't know which. Therefore we calculate the errors on the smoothed fit, increase them drastically by some factor from 3 to 15 according to how things look on the graphs, and feed the results for every partial wave back into the phase shift analyses at each energy as if they were a new sort of data. They represent the pull on this energy of the solutions at neighboring energies. This second round of phase shift analysis will now be overdetermined, and so we can ignore those smoothed predictions which really contradict the experiments and still be helped by the acceptable ones. The resulting phases will now be smoother and have smaller errors, so that the next energy-dependent smoothing can use more variables and get better fits. Iterating this process gradually drives the experimental phases and the smoothed fits toward each other. In practice, if one turns the screw too far the machine breaks and the experimental phase shift analyses cease to be good fits. This is just like energy-dependent fits where systematic errors in the data and inadequacies in the parametrization prevent χ^2 from reaching its expected value. The difference is that we have effectively something like 400 variables instead of 40.

In the resulting solutions, the errors at the highest energies are about the same as the original experimental ones, but they get relatively smaller as the energy decreases, which is physically reasonable, while useful information can be obtained even from energies where the data is too incomplete to allow an unconstrained analysis.

You will notice that I said nothing about the actual smoothing formulas. We used parametrizations that satisfy partial-wave dispersion relations, as I will describe later, but I have no strong feelings against generalized effective range expansions, Breit-Wigner or other formulas provided the number of variables is sufficient.

So far only the CERN analysis has succeeded in reducing the phases to reasonable smoothness, and only the CERN group have ventured to announce new resonances of low elasticity. These things are related. Steiner speculated in his paper[1] that perhaps the wobbles in the Saclay and Berkeley phases would go away if they tried a little harder. We found that the price for making the wobbles go away is precisely the introduction of new resonances, and it is this fact that we alone have been able to get rid of the wobbles that constitutes the real evidence for our resonances.

3. Dispersion Relations

a. Derivation of Backward Dispersion Relations

So far only one-dimensional dispersion relations in the energy have found much use phenomenologically. The main types are forward, backward, and partial wave. I shall consider the backward in most detail, because they are the least known[14] and are not being discussed elsewhere in this volume. To avoid complex singularities, we write them in k^2 (the square of the center-of-mass momentum) rather than s. The easiest derivation starts from the one-dimensional Cini-Fubini form[15] of the Mandelstam representation

$$A^{(\pm)}(s, t) = \frac{1}{\pi} \int_{(M+\mu)^2}^{\infty} ds' \alpha^{(\pm)}(s', t) \left\{ \frac{1}{s' - s} \pm \frac{1}{s' - \bar{s}} \right\}$$
$$+ \frac{1}{\pi} \int_{4\mu^2}^{\infty} dt' \frac{a^{(\pm)}(s, t')}{t' - t} + C_A^{(\pm)}, \tag{10}$$

$$B^{(\pm)}(s, t) = G_r^2 \left(\frac{1}{M^2 - s} \mp \frac{1}{M^2 - \bar{s}} \right)$$
$$+ \frac{1}{\pi} \int_{(M+\mu)^2}^{\infty} ds' \beta^{(\pm)}(s', t) \left\{ \frac{1}{s' - s} \mp \frac{1}{s' - \bar{s}} \right\} \tag{11}$$
$$+ \frac{1}{\pi} \int_{4\mu^2}^{\infty} dt' \frac{b^{(\pm)}(s, t')}{t' - t} + C_B^{(\pm)}$$

where G_r is the πN coupling constant

$$f^2 = (1/4\pi)(G_r \mu/2M)^2 \tag{12}$$

In the backward direction ($\cos \theta = -1$) *only*, the following identities [(13)–(16)] hold

$$t = -4k^2 \tag{13}$$

$$\bar{s} = (E - \omega)^2 = (M^2 - \mu^2)^2/s \tag{14}$$

$$\left\{\frac{1}{s' - s} + \frac{1}{s' - \bar{s}}\right\} ds' = \left(\frac{1}{E'\omega'} + \frac{1}{k'^2 - k^2}\right) dk'^2 \tag{15}$$

$$\left\{\frac{1}{s' - s} - \frac{1}{s' - \bar{s}}\right\} ds' = \frac{E\omega}{k'^2 - k^2} \cdot \frac{dk'^2}{E'\omega'}. \tag{16}$$

[Here $E = (M^2 + k^2)^{1/2}$, $\omega = (\mu^2 + k^2)^{1/2}$.] Substituting (15) and (16) into (10) and (11) enables us to write backward dispersion relations in k^2. Their high energy behavior will be dominated by baryon Regge poles[16] and therefore they should easily converge without subtractions which we thus omit. It is convenient to introduce the amplitudes

$$F^{(+)}(k^2) = (1/M)A^{(+)}(s, -4k^2) + (\omega/E)B^{(+)}(s, -4k^2) \tag{17}$$

$$F^{(-)}(k^2) = (E/M\omega)A^{(-)}(s, -4k^2) + B^{(-)}(s, -4k^2) \tag{18}$$

$$G^{(+)}(k^2) = (1/E\omega)B^{(+)}(s, -4k^2) \tag{19}$$

$$G^{(-)}(k^2) = B^{(-)}(s, -4k^2) \tag{20}$$

for each of which the backward dispersion relation becomes

$$\mathrm{Re}\, X(k^2) = \xi \bigg/ \left(\frac{\mu^4}{4M^2} - \mu^2 - k^2\right) + \frac{1}{\pi}\int_0^\infty \frac{dk'^2 \,\mathrm{Im}\, X(k'^2)}{k'^2 - k^2}$$
$$+ \frac{1}{\pi}\int_{-\infty}^{-\mu^2} \frac{dk'^2 \,\mathrm{Im}\, X(k'^2)}{k'^2 - k^2} \tag{21}$$

the Born term residues for the four amplitudes being

$$\begin{aligned}
F^{(+)} &: \xi = 4\pi f^2 \mu^2/M^2 & &= 0.2782f^2 \\
F^{(-)} &: \xi = 8\pi f^2(1 - \mu^2/2M^2) &&= 24.855f^2 \\
G^{(+)} &: \xi = 16\pi f^2/\mu^2 &&= 50.265f^2 \\
G^{(-)} &: \xi = 8\pi f^2(1 - \mu^2/2M^2) &&= 24.855f^2
\end{aligned} \tag{22}$$

if k^2 is in pion mass units. $F^{(\pm)}$ have the advantage of being directly related to the backward differential cross sections, which means that the errors on them will be much smaller than those on $A^{(+)}$ and $B^{(+)}$ separately, even after phase shift analysis. (Note that it is very important to take into account correlations among the phase shifts when evaluating any forward or backward dispersion relation.)

The interest of the backward dispersion relations lies in their close relation to the $\pi\pi$ interaction in the left-hand cut region $k^2 < -\mu^2$. Just as the left-hand cut of forward dispersion relations comes entirely from baryon exchange, so that of backward dispersion relations comes entirely from meson exchange, and they therefore give the cleanest way of separating this from πN scattering. First we expand the amplitudes (17)–(20) in terms of the helicity amplitudes[17] for the process $N\overline{N} \to \pi\pi$:

$$F^{(+)}(-t/4) = -(4\pi/Mp^2)f_+^{0}(t) - (20\pi/M)q^2f_+^{2}(t) + \cdots \qquad (23)$$

$$G^{(+)}(-t/4) = (60\pi/\sqrt{6})f_-^{2}(t) + \cdots \qquad (24)$$

$$F^{(-)}(-t/4) = (12\pi/M)f_+^{1}(t) + (28\pi/M)(pq)^2f_+^{3}(t) + \cdots \qquad (25)$$

$$G^{(-)}(-t/4) = 6\pi\sqrt{2}f_-^{1}(t) + 28\pi\sqrt{3}(pq)^2 f_-^{3}(t) + \cdots \qquad (26)$$

Here $p = (t/4 - M^2)^{1/2}$, $q = (t/4 - \mu^2)^{1/2}$, and $f_\pm^{J}(t)$ are the two helicity amplitudes for total spin J. The isospin will be 0 or 1 for J even or odd. It can be shown that (23)–(26) all remain finite at the $\pi\pi$ threshold $(q = 0)$ and at the $N\overline{N}$ threshold $(p = 0)$, the p^{-2} in the first term of (23) being cancelled by a P-wave centrifugal barrier factor.

Now the process $N\overline{N} \to \pi\pi$ has a big unphysical region between the $\pi\pi$ threshold $t = 4\mu^2$ and the $N\overline{N}$ threshold $t = 4M^2$. In the lower part of this unphysical region it follows from generalized unitarity[18] that the phase of each helicity amplitude will be the same (modulo 180°) as the corresponding $\pi\pi$ phase. Thus

$$f_+^{J}(t) = |f_+^{J}(t)| \exp[i\delta_J(t^{1/2}) + in\pi] \qquad (27)$$

where $\delta_J(m)$ are the $\pi\pi$ phases at total mass m for spin J and isospin 0 or 1. This equality holds up to the mass at which the process $2\pi \to 4\pi$ becomes important, i.e., certainly up to 558 MeV, and probably judging from peripheral experiments up to 900 MeV. Furthermore at sufficiently low $\pi\pi$ mass, the higher waves in (23)–(26) will be suppressed by the q^2 factors, so that from a knowledge of these amplitudes on their left-hand cuts we can obtain the $\pi\pi$ S_0, P_1, and D_0 phases at sufficiently low masses. Of course, we first have to reconstruct the amplitudes on the left-hand cut from their real and imaginary parts in the physical region, and I shall now describe the numerical method for this and other dispersion theoretic problems.

b. Numerical Evaluation of Dispersion Relations

The usual way of evaluating forward dispersion relations is to draw smooth curves through the experimental points for the imaginary parts, and then evaluate the dispersion integral for the real parts numerically. However, this method makes it impossible to investigate the effect of experimental errors, to which the singular integral will be very sensitive. This is even more important for non-forward dispersion relations, since the total cross sections are unusually well measured. The CERN evaluations of backward and partial wave dispersion relations therefore fit the experimental real and imaginary parts simultaneously by a series of complex functions chosen to satisfy the dispersion relations, each being multiplied by an adjustable real parameter.

Consider for example the right-hand cut of a partial-wave dispersion relation for the amplitude

$$f_{l\pm}{}^I = \frac{1}{2ik^{2l+1}}\,(\eta_{l\pm}{}^I \exp[2i\,\delta_{l\pm}{}^I] - 1) \tag{28}$$

Write the dispersion relation in

$$v = [s - (M + \mu)^2]/2M \tag{29}$$

so that the right-hand cut is $(0, \infty)$. It follows from unitarity and the existence of the centrifugal barrier that

$$\operatorname{Im} f_{l+}{}^I(v) = 0(v^{l+1/2}), \quad \text{as } v \to 0 \tag{30}$$

$$\operatorname{Im} f_{l+}{}^I(v) = 0(v^{-l-1/2}), \quad \text{as } v \to \infty \tag{31}$$

Now we introduce the new variable

$$x = (v_0 - v)/(v_0 + v) \tag{32}$$

where v_0 is an arbitrary positive constant, which transforms $v = (0, \infty)$ into $x = (1, -1)$. We then expand the imaginary part as a series of Gegenbauer polynomials in x, in order to ensure the threshold behavior (30)–(31):

$$\operatorname{Im} f_{l\pm}{}^I(v) = \sum_{n=1}^{\infty} a_n h_n{}^l(v) \tag{33}$$

where

$$h_n{}^l(v) = \left[\frac{2(v_0 v)^{1/2}}{v_0 + v}\right]^{2l+1} C_{n-1}^{l+1}\left(\frac{v_0 - v}{v_0 + v}\right) \tag{34}$$

For $l = 0$ there is a close connection with Fourier series which enables the convergence of this expansion to be proved under very general and physically plausible conditions on $\operatorname{Im} f_{l+}{}^I(v)$.[19] The point of this expansion is that the dispersion integral

$$g_n{}^l(v) = \frac{P}{\pi}\int_0^{\infty} \frac{dv'\,h_n{}^l(v')}{v' - v} \tag{35}$$

can be evaluated in closed form[20] in terms of hypergeometric functions

$$g_{2m}^{l}(v) = \binom{l - \frac{1}{2}}{l}[F(m, -m - l; \frac{1}{2} - l; 1 - x^2) - 1]$$

$$g_{2m+1}^{l}(v) = \binom{l - \frac{1}{2}}{l}[xF(m + 1, -m - l; \frac{1}{2} - l; 1 - x^2) + 1] \quad \text{for } v \geq 0$$

$$g_{n}^{l}(v) = 2^{-2l}\binom{2l + n}{l}[x - (x^2 - 1)^{1/2}]^{n} \tag{36}$$

$$\times F\{n, -l; n + l + 1; [x - (x^2 - 1)^{1/2}]^2\}$$

$$+ (-1)^{n+1}\binom{l - \frac{1}{2}}{l}, \quad \text{for } v \leq 0$$

So the contribution of the right-hand cut to the real part is

$$[\mathrm{Re}\, f_{l\pm}^{I}(v)]_{\mathrm{r.h.cut}} = \sum_{n=1}^{\infty} a_n g_n^{l}(v) \tag{37}$$

To evaluate the dispersion relation, we therefore cut off the series (33) and (37) at some finite n, add terms to (37) to represent left-hand cuts, subtractions, poles, etc., and determine the real parameters a_n by statistical fits to the experimental real and imaginary parts, taking account of correlations between them. Any adjustable parameters in the left-hand cut and other terms will be determined simultaneously. The number of terms required in these expansions can be decided by standard statistical tests, and the calculation of errors on the resulting smoothed fits is also standard.

The partial-wave amplitudes have been used here for illustration; the method applies equally to backward dispersion relations provided we choose the value of l in (34) such as to give the correct threshold behavior.

c. Evaluation of the Backward Dispersion Relations

To evaluate the backward dispersion relations, we used the real and imaginary parts calculated from our experimental phase shifts up to 2190 MeV mass. Some restrictions on $F^{(+)}$ at higher energies can be obtained from the experimental backward differential cross sections.[21] For the right-hand cut we first separated out the important P_{33} contribution, fitted the phase shift with an effective range expansion and then fitted its contributions to the various backward amplitudes (17)–(20) with a 20-term expansion of the (33)–(37) form with $v = k^2$. Another 20-term expansion was used to represent the rest of the right-hand cut contribution. f^2 in (22) was varied, but with the value[22]

$$f^2 = 0.0822 \pm 0.0018 \tag{38}$$

included as data in the fit.* For the left-hand cut we have tried either pole terms to represent the exchange of meson resonances, or else expansions of the same type [(30), (37)] with $v = -\mu^2 - k^2$ and the appropriate threshold behavior. These latter enable the $\pi\pi$ phase shifts to be deduced by (23)–(27).

If the helicity amplitudes are represented by pole terms

$$f_{\pm}^{J}(t) \approx \frac{N_{\pm}^{J}}{m_J^2 - t} + \text{r.h.cut}, \qquad (39)$$

then the masses and coupling constants of the meson resonances required to fit the backward dispersion relations are shown in Table I. The masses

TABLE I

Particle	Helicity state	Mass (MeV)	Coupling $N_{\pm}^{J}$ ($\mu = 1$)
σ	f_{+}^{0}	437	237 ± 15
ρ	f_{+}^{1}	591	5.36 ± 0.14
ρ	f_{-}^{1}	591	4.05 ± 0.61
f^{0}	f_{-}^{2}	1253	-0.18 ± 0.21

were fixed during the fits, but different values were tried. In particular the ρ has to have less than its physical mass in the one-pole approximation. A very large and close σ pole is required for the $F^{(+)}$ dispersion relation. Allowing for the different effective mass, the ρ coupling constants are reasonably close to Hamilton's values,[23] Unlike other people[24] we find no clear evidence for any spin-flip coupling of the f^0.

Combining with the σ coupling to the nucleon deduced from NN scattering[25] gives

$$g_{\sigma NN} \cdot g_{\sigma\pi\pi} = 5.34 \pm 0.13$$
$$(g_{\sigma NN})^2 = 3.05 \qquad (40)$$

which would correspond to a width for the σ of 480 MeV. When the σ is not represented by a pole, but instead by an expansion ((33), (37)), the resulting $\pi\pi$ S_0 phase deduced from (23) and (27) invariably turns out to have a very broad ($\Gamma > 200$ MeV) resonance (even without using the NN

*The $F^{(+)}$ dispersion relation is not sensitive to f^2 (see eq. (22)). The $F^{(-)}$ dispersion relation tended to give a lower value, as do recent evaluations of the B forward dispersion relation (G. Höhler, private communication) and pp phase shift analyses [M. H. MacGregor, R. A. Arndt, R. M. Wright] Phys. Rev., 169, 1128 (1968). I now think that f^2 could well be as low as 0.076.

constraint). This was laughed at when we first proposed it,[21] but the peripheral experiments, which were then thought to contradict a broad ππ S-wave resonance, have since changed their tune.[26] It seems to me that having been right when the rest of the world was wrong is a good check of our extrapolation procedure.

More recently we have proceeded by taking possible forms for $ππ$ S_0 phase shifts, chosen for their agreement with other evidence, putting 10% errors on them, and then fitting them as data simultaneously with the $πN$ information to see what happened. We used 10 terms for the left-hand expansions, which is enough to reproduce a great variety of $ππ$ phases (the right-hand cut needs more, but this is to fit fine detail rather than general features).

Figure 1 shows Re $F^{(+)}$ in the low energy region after subtraction of the nucleon and 33 resonance terms, but with the other right-hand cut contributions left in. The data continue for several page-widths to the right. The horrible fit at the bottom, which misses several points by over 100 standard deviations, corresponds to no $ππ$ S_0 wave. (Incidentally, a decreasing phase would give an even worse fit than a zero one.) The top dotted line is the form calculated by Hamilton et al.[23] five years ago, having a $ππ$ S_0 scattering length $\sim 1\mu^{-1}$ but no resonance. This is only qualitatively correct as the data stands today. The two reasonable fits correspond to a $σ$ pole at mass 437 MeV with the above coupling, and to either of the two explicit forms for the $ππ$ S_0 phase shown in Fig. 2. (Their $F^{(+)}$ predictions are indistinguishable on the graph.) Before leaving Fig. 1, note that $πN$ scattering lengths. The open circle is Hamilton's value.[27] This would make the curve curl over at the top. The only way of getting such an effect from the dispersion relation is to have the $ππ$ S_0 phase start negative and then cross zero, as Hamilton himself found.[23] This would violently contradict current interpretations of K_{e4} decay.[28] The black square is from our new $πN$ S-wave scattering lengths

$$a_3 = -0.0700 \pm 0.0054$$
$$a_1 = 0.1957 \pm 0.0111 \tag{41}$$

determined from partial-wave dispersion relation fits, with no prejudice about the $ππ$ interaction. The curve now goes up smoothly and as a result the negative $ππ$ scattering length has disappeared. The $πN$ and $ππ$ scattering lengths are thus very closely correlated in these fits. Figure 2 shows the actual $ππ$ S_0 phases corresponding to the "broad $ε + f^0$" line in Fig. 1. Both have resonances, of course. The $ππ$ scattering lengths are 0.461 and 0.287, which are not much bigger than Weinberg's, even though the fit was constrained to the K_{e4} value $0.84 + 0.43$.[28] The solid line is very similar to the more reputable peripheral results.[26]

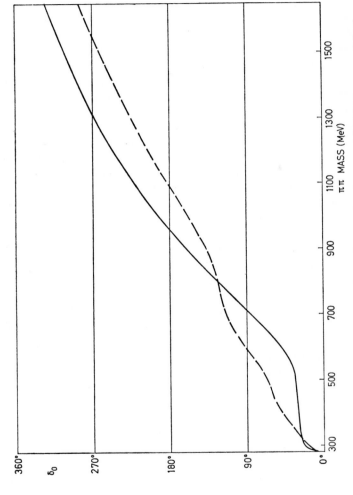

FIG. 1. Re $F^{(+)}$ in the physical region after subtraction of the nucleon and 33 resonance contributions.

40

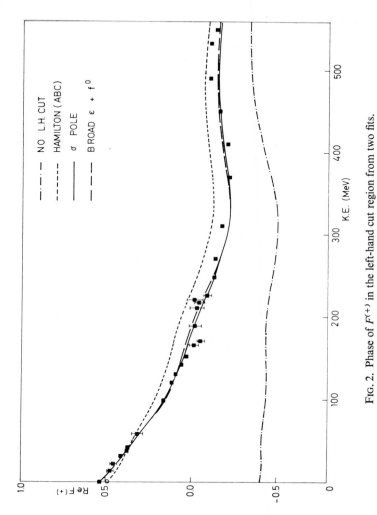

FIG. 2. Phase of $F^{(+)}$ in the left-hand cut region from two fits.

NO L.H. CUT

HAMILTON (ABC)

σ POLE

BROAD ϵ + f^0

K.E. (MeV)

Re $F^{(+)}$

41

We have also tried getting the $\pi\pi$ P_1 phase from the $F^{(-)}$ dispersion relation, instead of using a pole term. There is then no difficulty in giving the ρ its correct mass, but if a normal Breit-Wigner form is assumed then the width comes out much too big. We plan to try asymmetric forms with a CDD pole above the ρ, but have not yet done so.

The contribution of σ exchange to NN scattering is given by[25]

$$V_S{}^2(t) = \frac{1}{\pi}\int_{4\mu^2}^{\infty}\frac{dt'}{t'-t}\left[\frac{3M^2}{32\pi^2}\left(\frac{t'-4\mu^2}{t'}\right)^{1/2}|F^{(+)}(t')|^2\right] \tag{42}$$

We also evaluated this, using some of our πN backward dispersion relation fits. We found that the uncorrelated two-pion exchange (UTPE), i.e., the value of (42) with $F^{(+)}$ replaced by its Born + right-hand terms only, was much too small to explain the σ effect required by NN analyses. This disagrees with numerous statements in the literature. The contradiction is easily resolved: all previous authors[29] who calculated UTPE used the zero-width approximation for the dominant contribution of the 33 resonance to $F^{(+)}$. We found that when we did likewise it increased our UTPE by a factor of 11. This should be a warning to people who use this approximation in evaluating sum rules! On the other hand, our earlier fits[21] with a negative $\pi\pi$ scattering length would certainly make (42) orders of magnitude too large. It is very sensitive to the speed at which the $\pi\pi$ S_0 phase varies (the faster the larger), so that it would be more useful to fit the πN backward and NN forward dispersion relations simultaneously than to try to predict NN from πN.

d. Evaluation of Partial Wave Dispersion Relations

Because of the simplicity of their left-hand cuts, forward and backward dispersion relations are better for analysis—determination of coupling constants and $\pi\pi$ phase shifts. However, unitarity and the definite spins of resonances make partial wave dispersion relations more suited to synthesis—calculation of the physical effects in the s channel from these coupling constants, etc.

Instead of showing you what partial-wave dispersion relations look like, I shall discuss how we used them to smooth phase shifts and resolve ambiguities. The important thing to note is that we are interested in high energies and that the right-hand cut is then far more important than the left. Any physically reasonable left-hand cut must give a smooth contribution in the physical region so that if the phase shifts show a wobble this must either come from the right-hand cut or else be spurious. For this reason, the nearby left-hand cut was left as Hamilton and his disciples[30] calculated it, including his ABC contribution (large $\pi\pi$ scattering length) which we think should be replaced by a σ resonance. We did, however, check that this replacement changes the left-hand cut contribution to the phases by less than 2^0 at all energies except for the S waves. We assumed a 10% error on the nearby left-hand cut contribution calculated by DHL, with a separate 10% error on their ρ exchange term which becomes increasingly important at the higher energies. The distant left-hand cut was represented by three poles with arbitrary residues for the S waves, and two with the first centrifugal sum rule[31] imposed for P, D, and F waves. For G and H waves, we took the nearby left-hand cut contribution from unpublished calculations by Oades[32] and assumed that the distant part would be suppressed by the k^{-2l-1} factor.

For the right-hand cut, on the other hand, we were very sophisticated. The expansions (33) and (37) were used with more than 10 terms for almost all waves, and as many as 25 for some. Note that this method of evaluating partial-wave dispersion relations does not require any assumptions about the presence or absence of CDD poles. I consider this a great advantage over the N/D equations.

Besides experimental phases and elasticities with their correlations, we also used "pseudodata"

$$\eta = 1 \pm 0.001 \tag{43}$$

to impose elastic unitarity at low energies, and

$$\eta = 0.5 \pm 0.5 \tag{44}$$

to impose the unitarity limit at very high energies.

In P_{33} and P_{13}, we found definite evidence for the inadequacy of the DHL long-range forces. Their P_{13} is too negative below 300 MeV. Substitution of a σ for the ABC would certainly help here: in fact this is the only place where the difference is larger than the right-hand cut contribution. The trouble with P_{33} is that the fit to the phase with DHL forces disappears if unitarity is imposed to more than 2.5% in the 100 MeV region. Note that the Hamilton variational method, on the basis of which it was claimed that the DHL forces explained the P_{33} phase,[33] does not satisfy unitarity exactly. We do not yet know just what has to be modified. At higher energies the S_{11} and D_{13} dispersion relations fits have given trouble, though the latter is improved as a consequence of the new data. The agreement for all the other partial-wave dispersion relations is rather good, as can be seen from the graphs in my Heidelberg report.[3]

4. Conclusion

It is now more than 10 years since the first experiments on πN scattering in the region of the higher resonances. About 1961, most people wrote the subject off as dead. The more accurate measurements and phase shift analyses performed since then were achieved in the face of initial indifference and subsequent incredulity. This supposedly barren and exhausted field has turned out to be one of the most fruitful areas of high energy physics. The contrast with some of the fashionable theories of recent years is particularly striking.

I am grateful to Dr. G. L. Shaw for inviting me to present this paper, and to Drs. Riley, Yokosawa, Thresher, and Lundby for communicating their new data.

References

1. C. Johnson and H. Steiner, this volume (UCRL-18001).
2. A. Donnachie, R. G. Kirsopp, and C. Lovelace, *Phys. Letters*, **26B**, 161 (1968).
3. C. Lovelace, *Proc. Heidelberg Conf., 1967*, H. Filkuth, Ed., North-Holland, Publ. Co., Amsterdam, 1968, p. 79.
4. C. R. Cox *et al.*, Rutherford Report No. RHEL/M 137 (1967). C. R. Cox, K. S. Heard, J. C. Sleeman, P. J. Duke, R. E. Hill, W. R. Holley, D. P. Jones, F. C. Shoemaker, J. J. Thresher, and J. B. Warren.
5. P. Bareyre, C. Bricman, and G. Villet, *Phys. Rev.*, **165**, 1730 (1968).
6. A. S. Carroll *et al.*, *Phys. Rev. Letters*, **20**, 607 (1968). A. S. Carroll, J. Fischer, R. H. Phillips, C. L. Wang, A. Lundby, F. Lobrowicz, A. C. Melissinos, Y. Nagashima, and S. Tawksbury.
7. A. A. Carter, K. F. Riley, R. J. Tapper, D. V. Bugg, R. S. Gilmore, K. M. Knight, D. C. Salter, G. H. Stafford, E. J. N. Wilson, J. D. Davies, J. D. Dowell, P. M. Hattersley, R. J. Homer, and A. W. O'Dell, *Phys. Rev.*, **168**, 1457 (1968).
8. V. Barger and D. Cline, *Phys. Rev. Letters*, **20**, 607 (1968).
9. C. Michael, *Phys. Rev. Letters*, **21**, 93 (1966).
10. C. H. Johnson and H. Steiner, private communication.
11. W. C. Davidon, Argonne report ANL-5990 (1959); R. Fletcher and M. J. D. Powell, *Computer J.*, **6**, 163 (1964).
12. C. H. Johnson, UCRL-17683 (1967).
13. R. A. Arndt and M. H. MacGregor, *Phys. Rev.*, **141**, 873 (1966).
14. D. Atkinson, *Phys. Rev.*, **128**, 1908 (1962).
15. M. Cini and S. Fubini, *Ann. Physik.*, **10**, 352 (1960).
16. V. Barger and D. Cline, *Phys. Rev. Letters* **19**, 1504 (1967).
17. W. R. Frazer and J. R. Fulco, *Phys. Rev.*, **117**, 1603 (1960).
18. S. Mandelstam, *Phys. Rev. Letters*, **4**, 84 (1960).
19. I. Ciulli, S. Ciulli, and J. Fischer, *Nuovo Cimento*, **23**, 1129 (1962); C. Lovelace, *Nuovo Cimento*, **25**, 730 (1962); J. S. Levinger and R. F. Peierls, *Phys. Rev.*, **134**, 1341 (1964).

20. C. Lovelace, unpublished; A. Donnachie, in *Particle Interactions at High Energies*, T. W. Preist and L. L. J. Vick, Eds., Oliver and Boyd, Edinburgh, 1967, p. 376.
21. C. Lovelace, R. M. Heinz, and A. Donnachie, *Phys. Letters*, **22**, 332 (1966).
22. V. Samaranayake and W. S. Woolcock, *Phys. Rev. Letters*, **15**, 936 (1965).
23. J. Hamilton, P. Menotti, G. Oades, and L. Vick, *Phys. Rev.*, **128**, 1881 (1962).
24. N. Hiroshige, T. Ino, and M. Matsuda, *Progr. Theoret. Phys.*, **31**, 719 (1964); T. Ino, N. Hiroshige, M. Matsuda, and S. Sawada, *Progr. Theoret. Phys.*, **33**, 489 (1965); L. S. Azhgirei and V. I. Chizhikov, Dubna preprint El-3420 (1967).
25. A. Scotti and D. Y. Wong, *Phys. Rev.*, **138**, 145 (1965); R. A. Arndt, R. A. Bryan, and M. H. MacGregor, *Phys. Letters*, **21**, 314 (1966); S. Sawada, T. Veda, W. Watari, and M. Yonesawa, *Nuovo Cimento*, **49A**, 319 (1967); D. V. Bugg, *Nucl. Phys.*, **B5**, 29 (1968); further references are given in Ref. 3.
26. A. B. Clegg, *Phys. Rev.*, **163**, 1664 (1967); E. Malamud and P. E. Schlein, *Phys. Rev. Letters*, **19**, 1056 (1967); I. Butterworth, *Proc. Heidelberg Conf.* 1967, p. 64.
27. J. Hamilton, *Phys. Letters*, **20**, 687 (1966).
28. F. A. Behrends, A. Donnachie, and G. Oades, *Nucl. Phys.*, **B3**, 569 (1967).
29. D. Amati, E. Leader, and B. Vitale, *Phys. Rev.*, **130**, 750 (1963); W. N. Cottingham and R. Vinh Mau, *Phys, Rev.*, **130**, 735 (1963); J. W. Durso and P. Signell, *Phys. Rev.*, **135**, B1057 (1964); S. Furuichi and W. Watari, *Progr. Theoret. Phys.*, **34**, 594 (1965); **36**, 348 (1966); S. Furuichi and K. Watanabe, *Progr. Theoret. Phys.*, **35**, 408 (1966); J. W. Durso, *Phys. Rev.*, **149**, 1234 (1966).
30. A. Donnachie, J. Hamilton, and A. T. Lea, *Phys. Rev.*, **135**, B515 (1964).
31. A. Donnachie and J. Hamilton, *Phys. Rev.*, **138**, B678 (1965).
32. G. Oades, private communication.
33. A. Donnachie and J. Hamilton, *Phys. Rev.*, **133**, B1053 (1964).

N^* Studies via πp Inelastic Reactions

ARTHUR H. ROSENFELD and PAUL H. SÖDING†

*Department of Physics and Lawrence Radiation Laboratory,
University of California, Berkeley, California*

Introduction

Most of our knowledge of non-strange baryonic resonances has come from "elastic (scattering) phase shift analysis" (EPSA)

$$\pi p(\text{partial waves}) \to \pi p \qquad (1)$$

as discussed in this volume by Lovelace and Steiner.

In this paper we give:

1. A brief survey of inelastic cross sections (mainly via figures) and experiments (via two tables).

2. A summary of what very little extra insight (beyond that from EPSA) has come so far from the inelastic final states of N^* resonances when these are formed in the s channel

$$\pi p \to N^* \to \pi\pi N, \, \eta N, \, K\Lambda, \ldots \qquad (2)$$

3. A discussion of what useful information (branching ratios and signs of amplitudes) may be expected in the next year or so, and their relation to $SU(3)$ classification.

We considered, but did not write, an additional section on the production of N^*'s in reactions like $\pi p \to \pi N^*$, $pp \to N^*p$. However, we concluded that these experiments, while valuable as studies of the production process, are not yet very illuminating in our context. They do yield significant bumps[1] at N^*(1400, 1512, 1688, ...), and Δ(1236 and 1920), and also a bump in the $\pi^+\pi^+p$ spectrum at 1560 MeV.[2] However, the N^*(1400) bump seems to involve constructive interference of both the P_{11} resonance production and diffraction dissociation;[3] the N^*(1512 and 1688) bumps probably involve several unresolved resonances, and the $\pi^+\pi^+p$(1560) could well be a kinematic effect.[2]

† On leave from University of Hamburg.

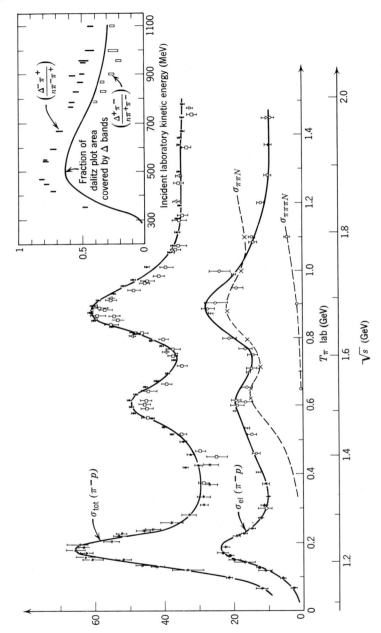

FIG. 1. $\pi^- p$ cross section. σ_{tot} and σ_{el} are from Ref. 4, $\sigma_{\pi\pi N}$ is from Ref. 5. The insert (fraction in Δ bands), from Ref. 5, is described in the text. For references on $\sigma_{\pi\pi\pi N}$, see Table I. Smooth curves have been drawn through the measured points.

I. Brief Pictorial Survey

First we summarize π^-p cross-section information[4,5] in Fig. 1. The difference between σ_{tot} and the sum of the lower curves is, of course, essentially made up by charge exchange. For $N^*_{3/2}$ resonances, $\sigma_{c.e.}/\sigma_{el} = \frac{2}{1}$; for $N^*_{1/2}, = \frac{1}{2}$. As we shall see in Fig. 4, the most abundant $\pi\pi N$ channel is $\pi^+\pi^-n$, then comes $\pi^-\pi^0p$, then $\pi^0\pi^0n$.

The Fig. 1 insert is a compact way of indicating the energy dependence of the population of the $\Delta(1236)$ bands of the $\pi^+\pi^-n$ Dalitz plot. The bars, solid and open, are experimental ratios; the curve, for comparison, shows where the *sum* of the two Δ bands would fall if the Dalitz plot population were uniform.

A typical $\pi^+\pi^-n$ Dalitz plot at $T_\pi = 900$ MeV ($s^{1/2} = 1688$ MeV) is shown in Fig. 2. Half the dots fall in the $\Delta^-(1238)$ band, even though it occupies only 15–20% of the area. The Δ^+ band shows no enhancement ($n\pi^+$ is much less strongly coupled to $I = \frac{3}{2}$). These facts are noted in the

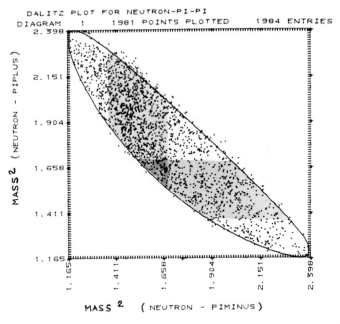

FIG. 2. Dalitz plot of ~ 2000 events of the reaction $\pi^-p \to \pi^+\pi^-n$ at $T_\pi = 900$ MeV ($s^{1/2} = 1688$ MeV). From LRL-SLAC collaboration (unpublished).

insert to Fig. 1 as a solid bar at 50% population for $T_\pi = 900$ MeV, an open bar at 25%, and a solid line at about 36% which represents the sum of the fractional areas of the two bands. The insert shows that $\pi^+\Delta^-$ is the dominant final state in $\pi^-p \to \pi\pi N$ over a large energy range. This should simplify the partial-wave analysis.

Figure 3 surveys π^+p cross sections.[6] Most of the comments that we made for π^- in Fig. 1 also apply to Fig. 3. Here there are only two charge states, $\pi^+\pi^0p$ and $\pi^+\pi^+n$.

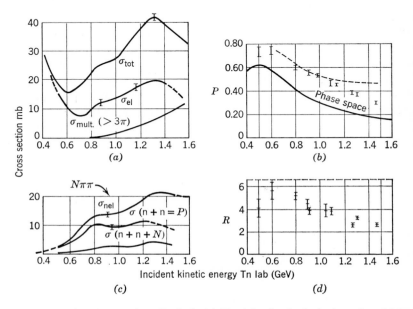

FIG. 3. π^+p cross sections (Ref. 6). (a) Total (σ_{tot}), elastic (σ_{el}), and multipion-production (σ_{mult}) cross sections for π^+p interaction from 0.4 to 1.4 GeV. (b) Total inelastic (σ_{inel}) and single-pion-production ($\sigma_{\pi^+p_\pi^0}$ and $\sigma_{\pi^+\pi^+n}$) cross sections. (c) Ratio P of the number of events in the isobar mass ($p\pi^+$) region (1.23 ± 0.07 GeV/c^2) to the total number of $\pi^+p\pi^0$ events compared with phase space (continuous (line) and isobar-model (dashed line) predictions. (d) Ratio R of the cross section $\sigma_{\pi^+p\pi0}$ to the cross section $\sigma_{\pi^+\pi^+n}$, and the isobar-model prediction (dashed line).

Figure 4 apportions the total $\sigma(\pi\pi N)$ of Fig. 1 (π^-p) and Fig. 3 (π^+p) into the individual charge channels,[7] and gives some measure of the consistency of the experimental data. Note the dominance of the two channels ($\pi^+\pi^-n$ and $\pi^+\pi^0p$) where we expect, and find, large Δ-band enhancements. To expand on this point we write down here Eqs. (3^-) and (3^+), which are just a display of some products of squares of Clebsch-Gordan coefficients.

We assume that the three $(\pi\pi N)^0$ channels come from a pure $I = \frac{1}{2}$ ampli-tude $\pi^- p \to N_{1/2}^{*0} \to \pi\Delta$. Then the resulting relative populations of the Δ bands are:

$$\pi^+ n\pi^- \quad : \pi^0 p\pi^- \quad : \pi^0 n\pi^0 \qquad (3^-)$$

$$1 + 9 = 10 : 2 + 2 = 4 : 4$$

Similarly for $\pi^+ p \to N_{3/2}^{*++} \to \pi\Delta$, we have

$$\pi^+ p\pi^0 \quad : \pi^+ n\pi^+ \qquad (3^+)$$

$$9 + 4 = 13 : 2$$

The experimental cross sections of Fig. 4 are in fact ordered in size in just the sequence given by Eqs. (3).

So far our pictorial survey has been too crude to mention partial waves; however, from EPSA one can of course predict the size of the inelastic partial waves; in fact Eqs. (4) show that σ_{inel} is just proportional to the hatched area on the Argand plot of Fig. 5. To see this, just compare[8]

$$\sigma_{\text{el}} = 4\pi\lambda^2 \left(J + \frac{1}{2}\right)|T|^2 = 4\pi\lambda^2 \left(J + \frac{1}{2}\right)\left|\frac{\eta \exp(2i\delta) - 1}{2i}\right|^2 \quad \text{(elastic)}$$

$$(4a)$$

$$\sigma_{\text{inel}} = \pi\lambda^2 \left(J + \frac{1}{2}\right)(1 - \eta^2) = 4\pi\lambda^2 \left(J + \frac{1}{2}\right)\left[\left(\frac{1}{2}\right)^2 - \left(\frac{\eta}{2}\right)^2\right] \quad \text{(inelastic)}$$

$$(4b)$$

Next we turn to the available experimental data, which we list in two tables:

Table I $(\pi^- p \to \pi\pi N)$. Two dozen small hydrogen bubble chamber experiments have been published with anywhere from a few hundred to 2000 inelastic events each, or a total of 23 000 events in all. Experiments now underway are listed in parentheses. A European effort is now aiming at 40 000 inelastic events per year, and an LRL-SLAC collaboration is comparable. (LRL-SLAC exposures are indicated as inserts of ~ 2000 inelastic events each.)

Table II $(\pi^+ p \to \pi\pi N)$. The published data are comparable with those of Table I, but there does not seem to be as much interest in large new runs, presumably because in this energy range there seem to be fewer Δ resonances to be sorted out, than $N_{1/2}^*$ resonances.

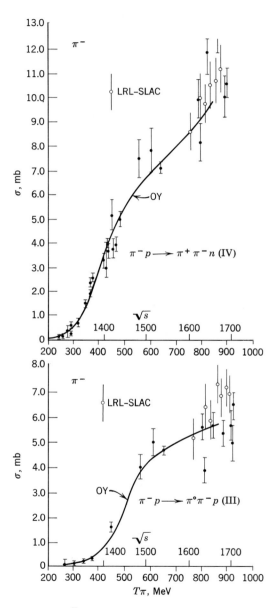

FIG. 4. Cross section for $\pi^\mp p \to \pi\pi N$, five charge states. Taken mainly from Ref. 7.

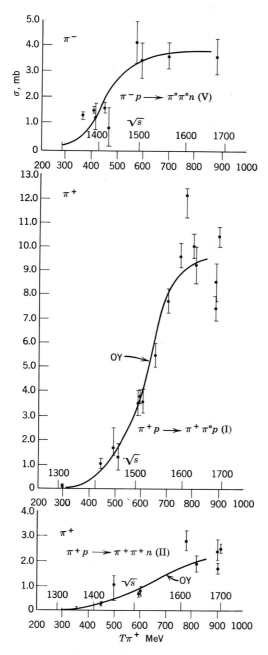

The solid curves are the results of the fit of Olsson and Yodh, Ref. 10, as discussed in Section II.

TABLE I
Some π^-p Inelastic Hydrogen Bubble Chamber Experiments[a]

$(s)^{1/2}$ (MeV)	T_π (MeV)	No. events of		Reference
		$\pi^-\pi^0p$	$\pi^+\pi^-n$	
		π^-p — $\pi\pi N$		
1305	290		250	Batusov et al., *Soviet Phys.*, *JETP*, **13**, 320 (1961)
	330		(3500)	Oxford (145 K pictures taken)
	360		573	Kirz et al., *Phys. Rev.*, **130**, 2481 (1963)
	370		(3000)	Oxford (60 K pictures taken)
1390	410	(500)	(1600)	Saclay (60 K pictures taken)
	420		(4500)	Oxford
1400	430		450	Kirz et al., *Phys. Rev.*, **130**, 2481 (1963)
1416	450	100	325	Poirier et al., *Phys. Rev.*, **148**, 1311 (1966)
	460		450	Kirz et al., *Phys. Rev.*, **130**, 2481 (1963)
2K→	480		329	Kirz et al., *Phys. Rev.*, **130**, 2481 (1963)
1450	490	(1800)	(5000)	Saclay (60 K pictures taken)
1450	500	(1500)	(4000)	Oxford (50 K pictures taken)
1480	550	(2300)	(3800)	Saclay
	555		450	Kirz et al., *Phys. Rev.*, **130**, 2481 (1963)
	558	441	833	Burnstein et al., *Phys. Rev.*, **137**, B1044 (1965)
2K→	604	1359	1970	Vittitoe et al., *Phys. Rev.*, **135**, B232 (1964)
	605		450	Kirz et al., *Phys. Rev.*, **130**, 2481 (1963)
	646	1049	1609	Oliver et al., *Phys. Rev.*, **147**, 932 (1966)
1525	620	(3600)	(5000)	Saclay (60 K pictures taken)
	650	538	777	Femino et al., *Nuovo Cimento*, **52**A, 892 (1967)
	673		450	Kirz et al., *Phys. Rev.*, **130**, 2481 (1963)
2K→	765	897	560	Crittenden et al., Sienna I, p. 116 (1963)
2K→	775	833	1600	Bertanza et al., *Nuovo Cimento*, **44**A, 712 (1966)

(continued)

TABLE I (*continued*)

$(s)^{1/2}$ (MeV)		T_π (MeV)	No. events of		Reference
			$\pi^-\pi^0 p$	$\pi^+\pi^- n$	
			$\pi^- p$—$\pi\pi N$		
		780		450	Kirz et al., *Phys. Rev.*, **130**, 2481 (1963)
		790	466	942	Cason et al., *Phys. Rev.*, **150**, 1134 (1966)
	2K→ 2K→	800	~200	~300	Gensollen et al., Sienna I, p. 84 (1963)
	2K→ 2K→	830	414	881	Cason et al., *Phys. Rev.*, **150**, 1134 (1966)
	2K→ 2K→	870	493	997	Cason et al., *Phys. Rev.*, **150**, 1134 (1966)
1688		900	671	1183	Gensollen et al., Sienna I, p. 84 (1963)
	2K→ 2K→	905	216	354	Pickup et al., *Phys., Rev.* **132**, 1819 (1963)
	2K→	960	262	384	Pickup et al., *Phys. Rev.*, **132**, 1819 (1963)
1800		1100	263	436	Pickup et al., *Phys. Rev.*, **132**, 1819 (1963)
		1100	(15 000)		Manning and Smith (LRL, 1968 to be publ.) (from $\pi^+ d$—$pp\ \pi^+\pi^-$)
2300		2200	(25 000)		LRL-SLAC (total LRL-SLAC, ~50 000 inelastic events spread over a $(s)^{1/2}$ range of 500 MeV, so ~2000 events every 20 MeV)
Total analyzed:			~8000	~15 000	
			$\pi^- p \to K^0 \Lambda$		
1613		768 = threshold			
1688		900	(8500)		Anderson, Crawford, Doyle (LRL)

[a] Numbers of events enclosed in parentheses indicate experiments still in progress; for other comments, see text.

TABLE II

Some π^+p Inelastic Hydrogen Bubble Chamber Experiments

$s^{1/2}$ (MeV)	T_π (MeV)	Events in channel		Reference
		$\pi^+\pi^0p$	$\pi^+\pi^+n$	

$\pi^+p \to \pi\pi N$ $T = 180$ MeV (threshold) to 1100 MeV

$s^{1/2}$ (MeV)	T_π (MeV)	$\pi^+\pi^0p$	$\pi^+\pi^+n$	Reference
	357		213	Kirz et al., *Phys. Rev.*, **126**, 763 (1962)
1415	450	100	28	Poirier et al., *Phys. Rev.*, **148**, 1311 (1966)
	500	159	40	Debaisieux et al., *Nucl. Phys.* **63**, 273 (1965)
1510	600	418	75	Newcomb et al., *Phys. Rev.*, **132**, 1283 (1963)
	810	2200		Deler et al., to be published
	820	346	73	Barloutaud et al., *Nuovo Cimento*, **27**, 238 (1963)
1690	900	274	75	Barloutaud et al., *Nuovo Cimento*, **27**, 238 (1963)
	900	274	201	Gensollen et al., Sienna I, p. 84 (1963)
	900	2517	529	Metzger et al., UR-875-186
	910	846	209	Stonehill et al., *Rev. Mod. Phys.*, **34**, 503 (1962)
	980	4105		Tautfest and Willmann, Athens (Ohio), p. 421 (1965)

(*continued*

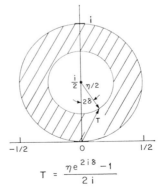

$$T = \frac{\eta e^{2i\delta} - 1}{2i}$$

FIG. 5. Argand plot for the elastic scattering amplitude.

TABLE II (*continued*)

$s^{1/2}$ (MeV)	T_π (MeV)	Events in channel		Reference
		$\pi^+\pi^0 p$	$\pi^+\pi^+ n$	

$\pi^+ p \to \pi\pi N$ $T = 180$ MeV (threshold) to 1100 MeV

$s^{1/2}$ (MeV)	T_π (MeV)	$\pi^+\pi^0 p$	$\pi^+\pi^+ n$	Reference
1760	1050	315	80	Barloutaud et al., *Nuovo Cimento*, 27, 238 (1963)
	1090	951	249	Stonehill et al., *Rev. Mod. Phys.*, 34, 503 (1962)
	1130	3860		Tautfest and Willmann, Athens (Ohio), p. 421 (1965)
	1260	1365	490	Stonehill et al., *Rev. Mod. Phys.*, 34 503 (1962)
1900	1300	3200		Deler et al., to be published (Saclay)
	1460	3150	1194	Daronian et al., *Nuovo Cimento*, 41A, 503 (1966)
	420			
	to 1070	$\frac{d\sigma}{d\omega}$ (π^0) only, counters		Detoeuf (1966)

$\pi^+ p \to K\Sigma$

$s^{1/2}$ (MeV)	T_π (MeV)			Reference
1690	900 = threshold			
1851	1207	(500)		Birge, Borreani, Kalmus, LRL (1968)
1896	1297	(700)		Birge, Borreani, Kalmus, LRL
2015	1546	(700)		Birge, Borreani, Kalmus, LRL

II. Inelastic Partial-Wave Analyses

To physicists familiar with the great successes of hydrogen bubble chamber s-channel resonance formation experiments in discovering and identifying $Y_0*(1520)$, $Y_1*(1660)$, $Y_1*(1770)$, and many others, it can be surprising that similar experiments with $\pi p \to \pi\pi N$ have yielded much less information.

The explanation is probably that until recently $K^- p$ experiments seemed more promising than those with πp, for several reasons:

(*a*) The $Y*$ resonances seemed to be narrower, hence easier to resolve, and more numerous.

(b) Copious inelastic final states like $\pi\Lambda$, $\pi\Sigma$, $\pi\Lambda(1520)$, are two-body, or quasi-two-body, and hence easier to analyze than $\pi\pi N$.

(c) The subsequent K or Y decays make it possible to observe even neutral particles, and yield extra data on the weak interaction, making each event more informative.

Anyway, whatever the reasons, the largest bubble chamber groups tended to build electrostatically separated K^- beams, collect and measure hundreds of thousands of events, and carry out detailed partial-wave fits to the data.

By contrast, πp experiments were considered less exciting and have really only become reputable with the advent of high-quality polarization data and the thorough phase shift analyses. Until recently, the πp inelastic channels tended to be studied by smaller, more dispersed groups, who measured fewer events and got less from each measurement.

Just as the πp *data* are less complete than the K^-p data, so correspondingly less effort has gone into *understanding* them.

At a given c.m. energy in a three-body final state (e.g., $\pi\pi N$, or $\pi\Delta$, where $\Delta \to \pi N$) the momenta are specified by four independent variables, which can be thought of as the two coordinates "internal" to the Dalitz plot, and two "external" angles, specifying, for example, the c.m. production angle θ_1 of particle 1, and the azimuth angle ϕ_{23} between the planes of production and decay of the diparticle (23). The Kp style of operating has been to write onto a data summary tape these four variables for each event, then make a partial-wave fit to all these data. Sometimes data from several experiments have then been pooled to extend the energy range and improve the statistics.

The πp analyses have been much less powerful. The early advocates of an "isobar model" were Sternheimer and Lindenbaum,[9] and later Olsson and Yodh.[10] Both papers fitted only mass and angular distributions. Let us discuss what fraction of the information one loses thereby. Since we are dealing with partial waves up to $J = \frac{7}{2}$, we expect Legendre polynomials up to $P_7(\cos\theta)$ to show up in an expansion. Hence a proper treatment would assign 10–20 bins to each variable; then in the space of all four variables we expect more than 10^4 significant bins. Instead, the current style is to fit at most the three (correlated) projections of the Dalitz plot and (omitting the correlations) about three more production angles, one for each particle. To our disappointment, it is hard to say how much information is thus lost, but it seems to be considerable.

We know of nobody so far who has bothered to collect any appreciable fraction of the total data available and make a complete fit. As we said above, now that the EPSA's have shown us there are many resonances under each bump, waiting for their $\pi\pi N$ modes to be disentangled, the field has become reputable, and you may expect things to change. But it will take a year. Meanwhile, and with apologies, let us discuss what little has been learned so far.

1. $I = \frac{1}{2}$, $\pi N \to \pi\pi N$

From the phase shift analyses of elastic πN scattering, it is known that in the energy region $T_\pi < 800$ MeV ($E_{c.m.} < 1630$ MeV), all partial waves with $J \leq \frac{5}{2}$ have sizable inelasticities. On the other hand, the amplitudes for $J \geq \frac{7}{2}$ are either elastic in this energy region, or just about to start becoming slightly inelastic. In Fig. 6 the partial-wave inelastic cross sections

$$\sigma_{inel} = \pi \lambdabar^2 (J + \tfrac{1}{2})(1 - \eta^2) \qquad (5)$$

(excluding elastic charge exchange) are shown for $J \leq \frac{3}{2}$. The values of the elasticity parameters $\eta = |S^{el}_{J,l}|$ are taken from the analyses of elastic πN scattering by different groups. Except for $J^P = \frac{3}{2}^+(P_{13})$, these inelastic cross sections show resonance behavior. The well-established $I = \frac{1}{2}$ resonances important for this energy region are listed in Table III.

TABLE III
Well-Established $N^*_{1/2}$ Resonances with Mass <1700 MeV

J^P	$\frac{1}{2}^+(P_{11})$	$\frac{1}{2}^-(S_{11})$		$\frac{3}{2}^-(D_{13})$	$\frac{5}{2}^-(D_{15})$	$\frac{5}{2}^+(F_{15})$
Mass (MeV)	1470	1550	1710	1525	1680	1690
T_π (MeV)	530	660	940	620	880	900
Γ_{mass} (MeV)	210	130	300	115	170	130
$\Gamma_{T\pi}$ (MeV)	320	220	550	190	300	230
σ_{inel} (mb)	6	$\sim 4^a$	1.8	~ 7	10	10
η (see Fig. 5)	0.35	~ 0.4	0.7	~ 0.2	0.2	0.2
Γ_{el}/Γ_{tot}	0.66	0.33	0.79	0.57	0.39	0.69

^{a} Mainly η production.

For all of these resonances except $S_{11}(1550)$, the dominant "inelastic" (i.e., non-πN) decay channel is expected to be $\pi\pi N$, whereas for the $S_{11}(1550)$ resonance, this decay mode is probably unimportant, the dominant "inelastic" mode being ηN in this case (see Fig. 6).

From analysis of the inelastic reaction $\pi N \to \pi\pi N$, one hopes to (i) separate the various reaction amplitudes according to their J^P values; (ii) further separate, for each J^P, the different interfering decay channels such as $\pi\Delta$, "σ" N, or "ρ" N. (Here "σ" stands for an $I = 0$, S-wave $\pi\pi$

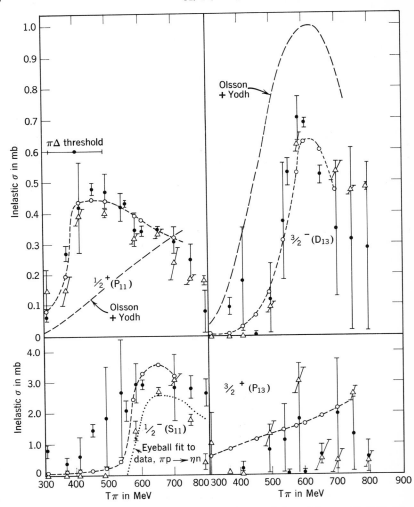

FIG. 6. Partial-wave inelastic cross sections for $I = \frac{1}{2}$ from Morgan (Ref. 7). The ordinate is *all* inelastic channels, therefore is only an upper limit for $\pi N \to \pi\pi N$. $\sigma_{\text{inelastic}}$ for $I = \frac{1}{2}$ partial waves, $S_{11}, P_{11}, P_{13}, D_{13}$. (-○-) Bransden et al., *Phys. Rev.*, **139B**, 1266 (1965). (△) Bareyre et al., *Phys. Letters*, **18**, 342 (1965). (⧗) Donnachie, Edinburgh Lectures (1966).

pair in an attractive but not necessarily resonant interaction, while "ρ" stands for a pion pair in the $I = 1$, P-wave state with a mass below or around the rho meson resonance mass. Note that the 1704-MeV threshold for production of pion pairs with a mass of $m_\rho = 765$ MeV is at $T_\pi = 925$ MeV.)

In contrast to the partial-wave analysis of elastic πN scattering, the difficulty with *three* particles in the final state is that the complete partial-wave expansion is *not unique*. For a given overall J^P one may, for example, expand the amplitude into a series of terms labeled by the quantum numbers shown in Fig. 7a. Using such an expansion [we call it the (12)3 expansion], one analyzes for definite angular momentum and isospin states of the (12) subsystem. Transition amplitudes into states with a definite angular momentum in, for example, the (23) subsystem would, in the (12)3 expansion, in general be represented by a large number of terms, and would not be easily recognizable. To simplify the problem it is usually assumed that, for each of the three pairs of particles in the final state, one or a few simple sets of quantum numbers dominate, due to some strong two-body final-state interactions in certain angular momentum and isospin states. For each two-particle subsystem, in addition to the few states assumed to be enhanced by the final-state interactions in this pair, other states will in general be superimposed as a consequence of the final-state interactions in the other pairs. In this way, one arrives at an isobar model, with interactions in each of the pairs (Fig. 7b). If one further specifies the dynamical form of the interactions in each of the pairs (by an energy-dependent two-particle scattering phase shift, or simply by a complex propagator or scattering length), one has a complete parametrization of the three-particle production amplitude at fixed total energy E. This may then be fitted to the experimental data.

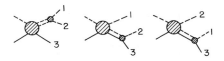

FIG. 7. (*a*) Possible partial-wave decomposition in the production of three particles. For $\pi\pi N$, only one has spin, but for generality we assume the nucleon may be either particle 1, 2, or 3, so we distinguish between, e.g., l_{12} and j_{12}. (*b*) The three amplitudes used in the isobar model.

The status of the phenomenology involved and the results from various attempts to analyze the published data on $\pi N \rightarrow \pi\pi N$ in the $I = \frac{1}{2}$ state have been recently reviewed by Morgan.[7] First, we discuss the methods, then the results.

Two different approaches have been used. *The first* of these, as described by Arnold and Uretsky,[11] is model-independent. One uses basically only information on the production angle distribution $d\sigma/d\Omega$, ignoring information on the specific properties of the decay of the intermediate state into the final $\pi\pi N$ state. One expands $d\sigma/d\Omega$ (which for a three-particle final state depends in general on two independent angles that can be specified in different ways[12]) into a series of, e.g., spherical harmonics:

$$d\sigma/d\Omega(\cos\Theta, \Phi) = \sum_{l,m} A_l^m Y_l^{m*}(\Theta, \Phi) \quad A_l^{-m} = (-)^m A_l^{m*}$$

The expansion coefficients A_l^m can be expressed as a sum over contributions from the various bilinear combinations of amplitudes corresponding to different J^P, in complete analogy with the Legendre polynomial expansion of elastic scattering differential cross sections.[13] In particular, with the choice (*a*) of the production angles Θ, Φ as defined in Ref. 12, and with M being the component of the total angular momentum $\mathbf{J}$ normal to the plane spanned by the three final-state c.m. momenta, the following rules apply[14]:

1. There can only be terms with $l + m$ even.[12a]
2. The interference between two partial waves with J, M and J', M' contributes only to A_l^m with $|J - J'| \leq l \leq J + J'$ and $m = M - M'$.
3. The absolute magnitude squared of a partial-wave amplitude of angular momentum J contributes to A_l^m with even $l \leq 2J - 1$.
4. The interference terms between two partial waves of even (odd) relative parity contribute only to A_l^m with $l =$ even (odd).
5. The amplitude labeled by J, M corresponds to a transition in a state with parity $(-)^{M-\mu}$, where μ is the component of the spin of the final nucleon normal to the plane of the three final-state c.m. momenta.

This expansion then allows one easily to estimate the highest contributing angular momentum J_{max}, and, hopefully, to identify the behavior of particular partial waves from the behavior of those expansion coefficients to which these partial waves contribute. One can study the expansion coefficients (as a function of the total energy, for example) either integrated over the Dalitz plot, or for individual regions of the Dalitz plot. Arnold and Uretsky[11] have used this method to check consistency of $\pi\pi N$ data with the inelasticity determinations as plotted in Fig. 6. Also Roberts[15] has used $d\sigma/d\Omega$ data restricted to the Δ band of the Dalitz plot and checked to see whether decay from the different J^P channels into just the $\pi\Delta$ configuration saturates the total inelasticity known from EPSA.

The second approach consists in fitting an isobar model,[16-19] that gives information on both (i) and (ii) above. Thurnauer,[17] Olsson and Yodh,[10] and Morgan[7], analyzing the region $T_\pi \leq 800$ MeV, have used essentially similar models in which the full amplitude is the sum of the three " isobar " amplitudes of Fig. 7*b*. Their assumptions differ mainly in which partial-wave transitions they include. Olsson and Yodh[10] use $\pi\Delta$ and $\pi(\pi N)_{I=1/2, S\,\text{wave}}$ final states (i.e., they assume πN final-state interactions in $I = \frac{1}{2}$, $J^P = \frac{1}{2}^-$ "$(\pi N)_S$" and $I = \frac{3}{2}$, $J^P = \frac{3}{2}^+$ "Δ" states). Thurnauer[17] uses $\pi\Delta$ and "σ"N. Both assume only S-wave production of the " isobars." Their assumptions are not as different as they might seem since there is very strong overlap between S-wave $\pi(\pi N)_S$ and "σ"N final states, in each case all three particles being in relative S waves.

Morgan[7] in addition considers "ρ"N states (the "ρ" being described by the tail of a resonance pole at 750 MeV), and adds P-wave production of the isobars.

As mentioned above, the final states considered in the isobar model in general belong to *different* complete orthogonal sets of states, corresponding to the different possible coupling schemes $(\pi_1\pi_2)N$, $\pi_1(\pi_2 N)$, or $\pi_2(\pi_1 N)$. Therefore, Namyslowski et al.[18] choose a somewhat different approach. They do not just add the three separate *complete* amplitudes, as given by the isobar model for transitions into these various final states. Instead, they project all final states onto a common set of complete states, say the one corresponding to the $(\pi_1\pi_2)N$ coupling scheme, and then only add the contributions, from the different terms of the isobar model, to the amplitudes for transition into the *few lowest* states of this set (i.e., just the ones whose presence is actually indicated by the experimental data). While this procedure is not much more complicated than the straight isobar model approach, Namyslowski et al. suggest that it is more systematic, since one deals with just one set of amplitudes, defined in terms of one complete set of quantum numbers of the (initial and) final state.

Naturally, this approach can be extended beyond the isobar model. However, the present experimental data, entirely lacking in polarization measurements, leave the phase shift analyses badly underdetermined. Hence in the analysis at present one is forced to make simplifying assumptions, which amounts to using an isobar model. Whether the Namyslowski et al. approach to the isobar model is a more economical description of what is going on in $\pi N \to \pi\pi N$ than the straight (" naive ") isobar model, may soon be learned as more accurate data are subjected to phenomenological analysis.

Finally, one may, within the isobar model, use t and u channel exchange amplitudes to describe the contributions of the partial waves with higher J to isobar production in closed form (i.e., without having to introduce arbitrary parameters for each isobar production partial-wave amplitude). Thus ρ exchange is expected to be important for $\pi N \to \pi \Delta$, and pion exchange for $\pi N \to \sigma N$ or $\pi N \to \rho N$.

The results from the published analyses are summarized in Fig. 8.

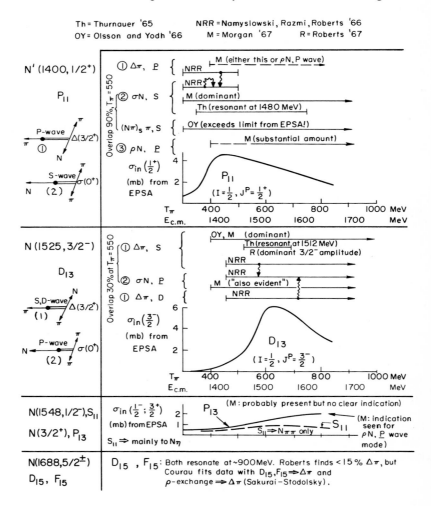

FIG. 8. Summary of partial waves for $I = \frac{1}{2}$, $\pi N \to \pi \pi N$, from various published analyses (Refs. 7, 10, 15, 17, 18, 21). The horizontal bars show the energy regions in which the partial waves listed on the left-hand side are present, according to the various analyses. Wiggly arrows symbolize the projections made by Namyslowski et al. (NRR), Ref. 18. For further explanations, see text.

(i) P_{11} : This partial wave is strongly seen in the $\pi N \to \pi\pi N$ reaction above $T_\pi = 300$ MeV. All analyses agree that there is $P_{11} \to$ "σ"N, S wave decay [which has 90% overlap[7] with $P_{11} \to \pi(\pi N)_{S\,\text{wave}}$, S wave, integrated over the Dalitz plot]. In addition, there are indications for $P_{11} \to \pi\Delta$, P wave, and/or $P_{11} \to$ "ρ"N, P wave. For $T_\pi \geq 550$ MeV, Morgan's analysis[7] suggests a large $P_{11} \to$ "ρ"N, P-wave amplitude, possibly resonant. In fact the CERN analysis[20] of elastic πN scattering suggests a second, very broad P_{11} resonance at $E_{\text{c.m.}} = 1750$ MeV ($T_\pi = 1010$ MeV).

(ii) D_{13} : The inelastic decay of this resonance ($\Gamma_{\text{in}}/\Gamma_{\text{tot}} \approx 40\%$) is clearly established to be dominantly $D_{13} \to \pi\Delta$, S wave. Only Namyslowski et al.[18] assume an equally large contribution of $D_{13} \to \pi\Delta$, D wave which seems somewhat unplausible in view of the centrifugal barriers involved. Morgan[7] finds in addition contribution from $D_{13} \to$ "σ"N, P wave.

(iii) P_{13} and S_{11} : Nothing is known yet about the magnitude and decay channels of the P_{13} and S_{11} into $\pi\pi N$. Also the decay modes of the D_{15} and F_{15} states (both resonant at $T_\pi \approx 900$ MeV) into $\pi\pi N$ have not yet been studied in detail. Roberts[15] finds very little ($<15\%$) $\pi\Delta$ final state, whereas Courau[21] fits the data with D_{15} and $F_{15} \to \pi\Delta$ plus ρ exchange leading to $\pi\Delta$; both authors, however, neglect interferences with other possible decay modes.

2. $I = \frac{3}{2}$, $\pi N \to \pi\pi N$

(a) Threshold ($T_\pi = 270$ MeV) to $T_\pi = 820$ MeV. The $\pi N \to \pi\pi N$ data in this region have been analyzed by Olsson and Yodh[10] and by Namyslowski et al.[18] Let us summarize their results and compare them with the results on inelasticities from the EPSA.

Olsson and Yodh make the most detailed fit to the data, using an isobar model. They find the dominant amplitude to be $D_{33} \to \pi\Delta$, S wave. To this they add a $P_{31} \to \pi(\pi N)_{I=1/2,\,S\,\text{wave}}$, S-wave amplitude to get the correct behavior of $\sigma(\pi^+ p \to \pi^+ \pi^0 p)/\sigma(\pi^+ p \to \pi^+ \pi^+ n)$. They choose this with the prejudice that S-wave states should dominate. To get the correct angular distributions in the $\pi^+ \pi^0 p$ channel, they have to add another amplitude from the same initial state, $P_{31} \to \pi\Delta$, P wave, as a "background term" (corresponding to about 10% contribution to $\sigma(\pi^+ p \to \pi^+ \pi^0 p)$ at $T_\pi = 600$ MeV). The phases and magnitudes of all three amplitudes are assumed independent of the total c.m. energy. The phases are relatively real, with the two P_{31} amplitudes being negative relative to the D_{33} amplitude. It is claimed that this model agrees with all the data available until 1966 below $T_\pi = 800$ MeV.

Namyslowski et al.,[18] on the other hand, compare with data at $T_\pi = 600$ MeV and $T_\pi = 820$ MeV only. They do not claim complete agreement; the worst discrepancy is with production angular distributions (where interferences between amplitudes for different J^P show up). In fact, their fits at $T_\pi = 600$ MeV to the angular distributions of the p and the π^0 in $\pi^+ p \to \pi^+ \pi^0 p$ are not as good as those of Yodh and Olsson.[10] They assume, at $T_\pi = 600$ MeV, a superposition of $D_{33} \to \pi\Delta$, S and D wave, and $S_{31} \to \pi\Delta$, D wave. At $T_\pi = 820$ MeV, they again take $D_{33} \to \pi\Delta$, S and D wave, but instead of S_{31}, they now add $P_{33} \to \pi\Delta$, P and F waves. They find that any inclusion of S_{31} destroys the fit at this energy.

Looking at the $I = \frac{3}{2}$ elasticity parameters η from the EPSA[20,23,24] one finds the following values:

T_π (MeV)	$s^{1/2}$ (MeV)	η (as defined in Fig. 5)					
		S_{31}	P_{31}	P_{33}	D_{33}	D_{35}	F_{35}
600	1512	0.9	1–0.9	≈ 0.95	0.9	1	1
820	1643	0.4 (res.)	1–0.8	≈ 0.8 (res ?)	0.7 (res ?)	0.8–0.95	≈ 0.9

For all the other partial waves, $\eta = 1$ in the energy region in question.

In one[20] of the EPSA, one finds $\eta(P_{31}) = 1$ for $T_\pi < 600$ MeV, which would be inconsistent with a sizable P_{31} contribution to $\pi N \to \pi\pi N$ in $T_\pi \leq 600$ MeV, as assumed by Yodh and Olsson.[10] It is well known, however, that the S_{31} amplitude goes through an inelastic resonance ($\eta = 0.4$) at $T_\pi = 820$ MeV, and P_{33} probably resonates at $T_\pi = 900$ MeV ($\eta = 0.7$). Therefore, one would expect to find a contribution from these partial waves in the $\pi N \to \pi\pi N$ channel. Regarding the Namyslowski et al.[18] analysis, it seems hard to understand that S_{31} should contribute at $T_\pi = 600$ MeV (i.e., below resonance), but not at the resonance itself where the inelasticity is largest.

(b) $T_\pi = 820$–1300 MeV. No detailed analysis has been reported in the literature. Only the production angular distributions have been discussed, and sometimes the analysis is restricted to $\pi\Delta$ final states. This selection becomes possible at these energies since the overlap region of the two Δ bands moves outside the Dalitz plot. However, as the energy increases, ρ production has to be taken into account ($>10\%$ for $T > 1000$ MeV).[6]

Kraybill et al.[6] expand the differential cross section for $\pi^+ p \to \pi^0 \Delta^{++}$ into a series of $P_l(\cos \theta)$ up to $l = 7$; they find no evidence for higher terms in this energy region. Deler et al.[22] choose a more general method, expanding the production angular distribution $d\sigma/d\Omega(\cos \Theta, \Phi)$ for $\pi^+ p \to \pi^+ \pi^0 p$ into a series of spherical harmonics $Y_l^m(\theta, \Phi)$. In both papers the expansion coefficients are then compared with theoretical expressions from a simple $\pi N \to \pi \Delta$ isobar model. The results are summarized below.

In the $T_\pi \approx 800$ MeV region, the presence of $P_{33} \to \pi\Delta$, P wave, and $D_{35} \to \pi\Delta$, D wave, are indicated, but other amplitudes are present as well. No firm conclusion is reached about the S_{31}, D_{33}, and P_{31} amplitudes, which are also expected to contribute. From the elastic phase shift analysis, P_{33} is known to probably resonate, and D_{35} to have already a sizeable inelasticity ($\eta = 0.95$ from the CERN,[20] $\eta = 0.8$ from the Saclay[23] phase shift analysis) in this energy region.

In the interval between $T_\pi = 900$ and 1200 MeV, the $F_{37} \to \pi\Delta$, F-wave amplitude seems to begin showing up, interfering with the $J^P = \frac{1}{2}^{\pm}$, $\frac{3}{2}^{\pm}$ amplitudes.

(c) $T_\pi \approx 1300$ MeV region ($E_{c.m.} \approx 1900$ MeV). The results are compatible with dominance of $F_{37} \to \pi\Delta$, F wave, which is near resonance there ($\eta \sim 0.4$). There could also be a $F_{35} \to \pi\Delta$, F-wave amplitude interfering with it; from the elastic scattering phase shift analysis, F_{35} is known to probably resonate at $T_\pi = 1330$ MeV with $\eta = 0.7$. However, no conclusion is possible about the size of the F_{35} amplitude from the inelastic reaction with the present data and analysis. The presence of further amplitudes of the same parity is indicated in the inelastic production angular distributions; these might be $P_{33} \to \pi\Delta$, P wave (which, however, has only $\eta \sim 0.9$ at this energy), or $P_{31} \to \pi\Delta$, P wave (which actually probably resonates[20] at $T_\pi = 1375$ MeV with $\eta \sim 0.4$). In addition, there seem to be some indications for the presence of amplitudes for negative parity also. The $D_{35} \to \pi\Delta$, D-wave amplitude seems to decrease above $T_\pi = 1000$ MeV and is found to be absent at $T_\pi = 1300$ MeV, although from the elastic scattering analysis its η is decreasing monotonically, with $\eta = 0.5–0.7$ at $T_\pi = 1300$ MeV, reaching a possible resonance at $T_\pi \approx 1400$ MeV (with $\Gamma \sim 310$ MeV according to the CERN EPSA[20]). These two considerations lead one to conclude that the D-wave $\pi\Delta$ decay of this resonance (if it exists) must be very small.

It might be mentioned that Courau[21] gets a rather satisfactory fit to the π^0 angular distribution in $\pi^+ p \to \pi^+ \pi^0 p$ at $T_\pi = 1300$ MeV, using *only* a resonating $F_{37} \to \pi\Delta$, F-wave amplitude and adding a term in the cross section that describes noninterfering Δ production via ρ exchange.

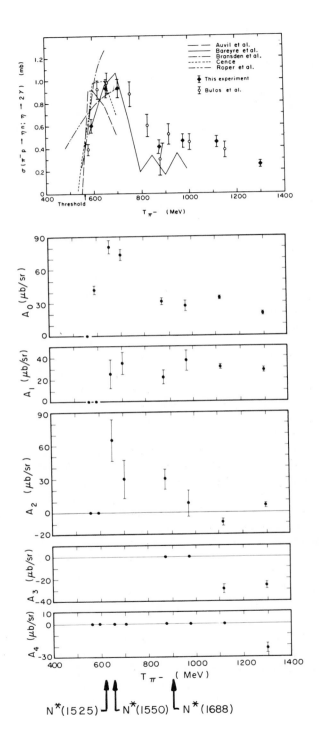

3. Other Channels: $\pi^- p \to \eta N, K\Lambda (I = \frac{1}{2})$

(a) $\pi N \to \eta N$. For the $\frac{1}{2}^-(S_{11})$ partial wave we have already shown in Fig. 6 that $\sigma_{\text{inel.}}(\frac{1}{2}^-)$ as determined from EPSA is within the errors accounted for by the sudden increase from threshold of $\sigma(\pi N \to \eta N)$. At the top of Fig. 9 we show the actual data[25] for $\sigma(\pi^- p \to \eta n)$, compared with $\sigma_{\text{inel}}(\frac{1}{2}^-)$ as predicted by various EPSA's. Then below are the partial cross sections, A_l, for the expansion $d\sigma/d\Omega = \Sigma A_l P_l(\cos \theta)$.

Below $T_\pi = 1000$ MeV ($E_{\text{c.m.}} = 1745$ MeV) only A_0, A_1, A_2 are different from zero. Besides S_{11}, other partial waves must be present; a combination of S_{11}, P_{11}, and D_{13} actually can explain the data[25] (although, by the Minami ambiguity, D_{13} could be replaced by P_{13}). At $T_\pi = 900$ MeV ($E_{\text{c.m.}} = 1688$ MeV) and $\frac{5}{2}^-(D_{15})$ and $\frac{5}{2}^+(F_{15})$ amplitudes are known to resonate and to be highly absorptive; nevertheless, no enhancement in $\sigma(\pi^- p \to \eta n)$ and no A_3, A_4 expansion coefficients are observed at this energy, which means that the relative decay rate into ηN of these resonances must be quite small ($<2.5\%$ and $<1.5\%$, respectively[30]).

It has been pointed out[25a] that the η production data are also quite consistent with the peak in $\sigma(\pi N \to \eta N)$ near threshold being dominated by a (new) P_{11} resonance at $E_{\text{c.m.}} = 1580$ MeV ($\Gamma = 130$ MeV, decay rate $\to \pi N$ 35%; $\to \eta N$ 30%), which must then be different from the well-established P_{11} resonance at $E_{\text{c.m.}} = 1470$ MeV. In the EPSA's, [20,23,24] no indication of such a resonance is seen; however, it seems that its existence can at present not be excluded on the basis of the elastic data alone. Measurement of the recoil nucleon polarization could resolve this ambiguity in the η production data.

(b) $\pi N \to K\Lambda$. Cross-section and angular distribution data[26] for $\pi^- p \to K^0 \Lambda$ between threshold (at $E_{\text{c.m.}} = 1615$ MeV) and $E_{\text{c.m.}} = 1750$ MeV are shown in Fig. 10. The large values of the A_1, A_2, and A_3 coefficients in the Legendre polynomial expansion of $d\sigma/d\Omega$ indicate considerable interference between at least two partial-wave amplitudes of different parity; from the absence of significant A_4 and A_5 coefficients at $E_{\text{c.m.}} = 1680$ MeV it appears, however, that the contributions from the resonant D_{15} (1680) and $F_{15}(1688)$ partial waves to the $K\Lambda$ channel are small. Upper limits on the $K\Lambda$ decay branching ratios of these resonances, as inferred[30] from the experimental limits on A_4 and A_5, are given in Table IV. Further analysis of these data is in progress.[26]

FIG. 9. η production cross section (top), and Legendre polynomial expansion coefficients of the η differential cross section, from Ref. 25. The cross-section values are for only the 34% of η which decay into $\gamma\gamma$. The η production cross section is compared with the S_{11} inelastic cross section predicted by various EPSA's.

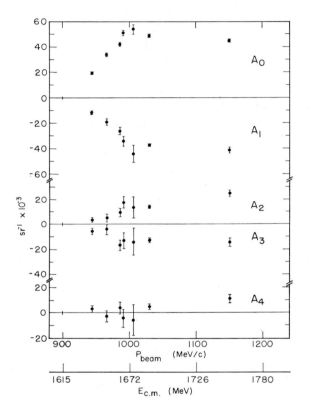

FIG. 10. Coefficients in the Legendre polynomial expansion $d\sigma/d\Omega = \lambda^2 \Sigma A_l P_l$.$(\cos \theta)$ of the differential cross section for the reaction $\pi^- p \to \Lambda K^0$. Note that θ is the c.m. angle between the momenta of π^- and Λ. No A_l with $l > 4$ were required for adequate fits in this momentum region. The total cross section for this reaction is given by $\sigma(\pi^- p \to \Lambda K^0) = 4\pi\lambda^2 A_0$ (from Ref. 26).

TABLE IV
Branching Fractions and Partial Widths for
Baryon* → 8 ⊗ 8 (from Ref. 30)

Mass and width	Mode	Branching fraction	Partial Γ
$J^P = \frac{1}{2}^-$			
$\Lambda(1405)$	$\Sigma\pi$	1.0	35
$N(1570)$	$N\pi$	0.3	39
$\Gamma = 130$	$N\eta$	0.7	91
$\Lambda(1670)^a$	$N\bar{K}$	0.06	1.1
$\Gamma = 18$	$\Lambda\eta$	0.94	16.9
$J^P = \frac{3}{2}^+$			
$\Delta(1236)$	$N\pi$	1.0	120
$\Sigma(1385)^b$	$\Lambda\pi$	0.86	30.1
$\Gamma = 35$	$\Sigma\pi$	0.14	4.9
$\Xi(1530)$	$\Xi\pi$	1.0	7.3
$J^P = \frac{3}{2}^-$			
$N(1530)^i$	$N\pi$	0.65	68
$\Gamma = 105$	$N\eta$		0.4
$\Sigma(1660)^c$	$\Lambda\pi$	0.10	5.0
$\Gamma = 50$	$\Sigma\pi$	0.67	33.5
	$N\bar{K}$	0.10	5.0
$\Lambda(1690)^d$	$\Sigma\pi$	0.46	18.4
$\Gamma = 40$	$N\bar{K}$	0.245	9.8
	$\Lambda\eta$	<0.027	<1.1
$\Xi(1815)^e$	$\Xi\pi$	0.10	1.6
$\Gamma = 16$	ΛK	0.65	10.4
	$\Sigma\bar{K}$	<0.02	<0.3
$\Lambda(1520)^j$	$\Sigma\pi$	0.51	8.2
$\Gamma = 16$	$N\bar{K}$	0.39	6.2
$J^P = \frac{5}{2}^+$			
$N(1688)^f$	$N\pi$	0.65	71.5
$\Gamma = 110$	$\Lambda\bar{K}$	<0.0013	<0.15
	$N\eta$	<0.015	<1.7
$\Lambda(1820)^g$	$\Sigma\pi$	0.12	10.2
$\Gamma = 85$	$N\bar{K}$	0.60	51
	$\Lambda\eta$	<0.014	<1.15

(*continued*)

TABLE IV (*continued*)

Mass and width	Mode	Branching fraction	Partial Γ
$\Sigma(1910)^g$ $\Gamma = 60$	$\Lambda\pi$	0.10	6
	$\Sigma\pi$	<0.01	<0.6
	$N\bar{K}$	0.08	4.8
$J^P = \frac{5}{2}^-$			
$N(1688)^f$ $\Gamma = 140$	$N\pi$	0.40	56
	ΛK	<0.016	<2.3
	$N\pi$	<0.025	<3.5
$\Sigma(1765)$ $\Gamma = 90$	$\Lambda\pi$	0.17	15.3
	$\Sigma\pi$	0.01	0.9
	$N\bar{K}$	0.50	45
	$\Sigma\pi$	<0.005	<0.5
$\Lambda(1827)^g$ $\Gamma = 75$	$\Sigma\pi$	0.23	17.2
	$N\bar{K}$	0.10	7.5
	$\Lambda\pi$	<0.08	<6.1
$\Xi(1933)$ $\Gamma = 140$	$\Xi\pi$	0.5	70
	$\Lambda\bar{K}$	0.5	70
$J^P = \frac{5}{2}^+$			
$\Delta(1920)$	$N\pi$	0.5	100
$\Sigma(2035)^h$ $\Gamma = 160$	$\Lambda\pi$	0.25	40
	$\Sigma\pi$	0.06	9.6
	$N\bar{K}$	0.16	25.6
	ΞK	<0.016	<2.6
$J^P = \frac{7}{2}^-$			
$\Lambda(2100)^h$ $\Gamma = 160$	$\Sigma\pi$	0.05	8.0
	$N\bar{K}$	0.29	46.4
	ΞK	0.01	1.6
	$\Lambda\eta$	<0.03	<4.8

[a] Branching fractions obtained on the two-channel assumption.

[b] Width and branching fractions from Ref. 9.

[c] We adopt slightly higher elasticity than that reported in Ref. 2. Even so, there is insufficient $\Sigma(1660)$ formed to accomodate the large $\Sigma\pi$ amplitude required by the analysis of Ref. 3 and in addition a comparably large rate into $\Lambda(1405)\pi$.

[d] Branching fractions for $\Sigma\pi$ and $N\bar{K}$ from Refs. 2 and 3. The upper limit for $\Lambda\eta$ comes from the measured cross section of 0.08 mb at $\frac{1}{2}\Gamma$ above resonance as reported by Berley et al.[16]

[e] Upper limit on the $\Sigma\bar{K}$ mode is extracted from Table 1 of the paper of Smith et al.[17] by comparison of the $\Sigma\bar{K}K$ and $\Lambda\bar{K}K$ reactions.

[f] The upper limits of the decay mode $N(1688) \to \Lambda K$ were extracted from the unpublished associated production data of Anderson et al.[18] The limits come from the absence of both A_4 and A_5 coefficients in the angular distributions of $\pi^- p \to \Lambda K^0$ in this momentum region. The coefficient A_5 is the more sensitive measure, and under the assumption that the two degenerate resonances of $J^P = \frac{5}{2}^+$ decay into ΛK in proportion to their probabilities of formation and to their respective centrifugal barriers, we obtain the limits listed. This ignores $SU(3)$ as a starting value. Figure 3c, d shows that another iteration keeping A_5 fixed would satisfy $SU(3)$. The upper limit on $N \to N\eta$ comes from assigning a maximum enhancement of 0.3 mb to the reaction $\pi^- p \to N\eta$ at 1688 MeV as derived from the work of Richards et al.[12]

[g] The $\Sigma\pi$ and $N\bar{K}$ rates are from Refs. 1-3. The upper limits on the $\Lambda\eta$ rates for $\Lambda(1820)$ and $\Lambda(1827)$ are obtained from the measured cross section of 0.2 mb.

[h] Preliminary estimate of the $\Sigma\pi$ mode from Barbaro-Galtieri.[20] The ΞK mode is estimated from the cross section for $K^- p \to \Xi K$ reported by Berge et al.[21]: the $\Lambda\eta$ upper limit is from Flatté and Wohl.[21]

[i] $N\eta$ partial width from the analysis of Davies and Moorhouse.[22]

[j] A more recent compilation by Yodh[23] gives for $\Lambda(1520)$ a lower branching ratio $\Gamma(\Sigma\pi)/\Gamma(N\bar{K}) = 0.42/0.47$, which indicates a greater need for singlet–octet mixing than shown in Fig. 1.

III. Signs and Magnitudes of Inelastic Amplitudes, $SU(3)$ Assignments

We have already suggested that successes in K^-p experiments may predict a bright future for πp studies. Figure 11 shows another sort of K^-p result,[27] this time unique to inelastic channels. It shows Argand plots for $Y^* \to \pi\Sigma$; we recognize resonant circles for two cases with $J^P = \frac{3}{2}^-$, $\Lambda(1690)$ and $\Sigma(1660)$ and for two more $\frac{5}{2}^+$ cases, $\Lambda(1815)$, $\Sigma(1910)$. The point we want to illustrate here is that two of the curves are plotted " up " (i.e., with $+i$ components) and two " down," i.e., the *signs* of the partial-wave amplitudes have been determined from the experimental angular distributions.

Let us pursue further the significance of these signs. Consider two final-state amplitudes of some resonance Y^*, produced by K^-p:

$$K^-p \xrightarrow{g_{K^-p}} Y^* \xrightarrow{g_{K^-p}} K^-p; \quad \langle K^-p|T^{JP}|K^-p\rangle \propto \underset{\text{positive definite}}{g_{K^-p}^2}; \qquad (6)$$

$$K^-p \xrightarrow{g_{K^-p}} Y^* \xrightarrow{g_{\pi\Sigma}} \pi\Sigma; \quad \langle \pi\Sigma|T^{JP}|K^-p\rangle \propto g_{K^-p}g_{\pi\Sigma} = \pm|g_{K^-p}g_{\pi\Sigma}| \quad (7)$$

In the first case (elastic scattering), the initial and final coupling constants are the same, and the partial-wave amplitude is necessarily positive. In the

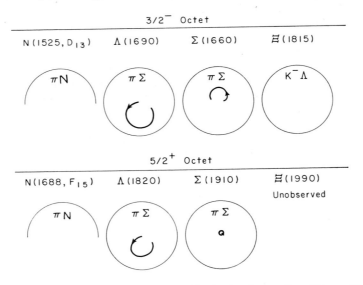

FIG. 11. Argand plots for partial-wave amplitudes in the reaction $K^-p \to \pi\Sigma$, from Ref. 27.

second (inelastic) case, we introduce a new coupling constant $g_{\pi\Sigma}$, and until we invoke $SU(3)$, all we know is that it is real. Of course we cannot even measure the sign of the amplitude until we let it interfere with another partial wave. But fortunately there are other resonances close enough in energy so that their two $\pi\Sigma$ modes overlap, and with adequate data the CERN-Heidelberg-Saclay group[27] has determined the relative signs, as shown in Fig. 11. Other determinations have been made by Kernan and Smart, who were among the first to study these signs.[28]

In the language of $SU(3)$, however, the sign of the amplitude [$\pi\Sigma$ in Eq. (7)] is no longer unknown, since there is of course only *one* real coupling constant g for reactions such as $Y^* \rightarrow 8 \times 1$ or 8×10. For the particular (but most common) case that Y^* is itself a member of an 8, and decays into 8×8, there can actually be two g's (although frequently one of them is zero); but even here, as we shall describe below, given a little additional data, the sign of the final amplitude is predicted and is a check on the $SU(3)$ assignment of the resonance Y^*. Further, though magnitudes may be hard to measure, signs are easier. In fact, about 10 such signs have now been reported, eight of which afford significant tests. The chance that some random $SU(3)$ assignment would pass all eight tests is then only 2^{-8}, so we see how valuable these signs can be.

To illustrate further the question of making $SU(3)$ assignments to the experimental data, we introduce Table IV and Fig. 12, both taken from Tripp et al.[30]

Note that Table IV is very weak on N^* and Δ inelastic branching fractions; it lists two ηN fractions, two ηN upper limits, and two $K\Lambda$ upper limits. With that limited information we conclude our N^* discussion; the rest of this section is only a summary of the status of $SU(3)$ assignments and a hint as to how better N^* data may be used in the future.

Table IV lists two possible 10's: the $\frac{3}{2}^+$ supermultiplet has three members whose branching fractions are related by $SU(3)$, the $\frac{7}{2}^+$ has two. In their Letter,[30] Tripp et al. show that the decay rates are compatible with the predictions of $SU(3)$ to within $\pm 30\%$. In addition, two singlet assignments [for $\Lambda(1520)$ and $\Lambda(2100)$] fit well if one allows for some singlet–octet mixing.

Finally Table IV lists four possible octets. Here, because of the complications that 8×8 couples to *two* 8's (one symmetric, 8_s, one antisymmetric, 8_a), the checking is not so straightforward. Tripp et al.[30] do it in an interesting graphical way in their Fig. 3, which is our Fig. 12. Since they had little room in their paper to explain their treatment of signs, we will expand on it here.

Since all the data are treated as partial widths, assume the resonance has been created with unit amplitude. For definiteness call the decaying

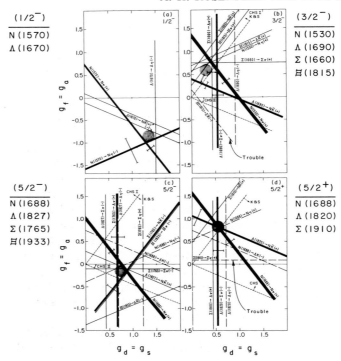

FIG. 12. Plots of g_s versus g_a from Tripp et al.[30]. Error bars on the heavy and medium lines correspond, respectively, to 25% and 50% uncertainty in the decay rates. Light lines are uncertain to a factor ≥ 2. Long-dashed lines denote upper limits. Short-dashed lines indicate the regions of the figures allowed by measurements of the relative signs of reaction amplitudes by Kernan and Smart and by the CERN-Heidelberg-Saclay collaboration. The sign affixed to each decay-rate line denotes the sign of the amplitude in Eq. (9). Shaded areas in each figure indicate the approximate values of g_s and g_a that seem to agree best with experiment, although other regions of each plot may still be acceptable. Wavy lines in Fig. 12b exhibit the displacements from the pure singlet and pure octet decay rates into $\Sigma\pi$ and $N\bar{K}$ due to a mixing angle of -16 deg, the magnitude suggested by the mass formula.

resonance Y^*, and remember it is part 8_s and part 8_a. Write the partial width Γ as

$$\Gamma = (c_s g_s + c_a g_a)^2 \times \text{Kinematics} \qquad (8)$$

where Kinematics stands for all kinematical and phase space factors, including barrier penetration factors B_l for orbital angular momentum l, i.e.,

$$\text{Kinematics} = B_l(p)(M_N/M_Y^*)p$$

and where c_s, c_a are isoscalar coefficients, thus for $\bar{K}N$,

$$\Gamma(\bar{K}N) = [c_s(\bar{K}N)g_s + c_a(\bar{K}N)g_a]^2 \times \text{Kinematics}$$

The square root of (8) yields an amplitude $A(\overline{K}N, \pi\Sigma, \cdots)$ as

$$A = c_s g_s + c_a g_a = \pm\,(\Gamma/\text{Kinematics})^{1/2}$$

If we plot $g_a \equiv y$ versus $g_s \equiv x$, we get straight lines,

$$y_\pm = \frac{\pm|A| - c_s x}{c_a} \tag{9}$$

Note that the greater the observed amplitude $|A|(\propto \Gamma^{1/2})$, the farther the two lines $y_\pm$ lie from the origin.

Since one overall sign is always undetermined, we chose $x = g_s$ to be >0, and plot only the right half plane. Satisfactory agreement among members of an octet is indicated by a common value of x and y for all decay modes. Shaded areas indicate where the lines cross. Tripp et al. discuss all four octets in turn, but we can see that for each octet the lines overlap within errors with one serious discrepancy in each case.

Finally we come to the question of the signs, which are shown in Fig. 12 as dashed lines with arrows indicating allowed regions. Figure 13

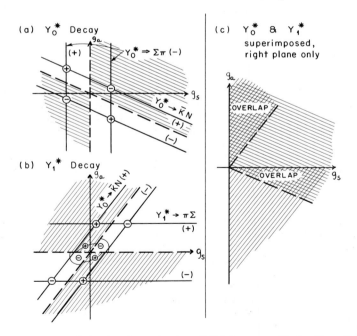

FIG. 13. Illustration of the sign determination of Y^* decay amplitudes. Signs indicated by $(+)$ or $(-)$ are those of the amplitude A in Eq. (9).

illustrates how some of these areas were constructed, using the fact of Fig. 11 that in the $\frac{3}{2}^-$ octet the $I = 1$ and $I = \pi\Sigma$ amplitudes are found to be of opposite sign.

Figure 13a shows the straight lines $y_\pm(\overline{K}N)$ from Eq. (9); y_+ is labeled $(+)$, etc. For the $\pi\Sigma$ decay mode, $c_a = 0$, giving vertical lines, $x_\pm = \pm |A|/c_s$, only one of which, x_-, is in the right half plane (since c_s happens to be $-(15/5)^{1/2}$. The line is labeled $(-)$. Since the reaction under discussion is $\overline{K}N \to Y^* \to \pi\Sigma$, the sign of $|\pi\Sigma\rangle$ is of course the sign of the product of $|\overline{K}N\rangle$ and $|\pi\Sigma\rangle$. We have written these products inside cirlces where the $\overline{K}N$ and $\pi\Sigma$ lines cross. The whole plane is then seen to be divided into four sectors, two of them $-$ (shaded), and two of them $+$ (unshaded).

Figure 13b repeats the reasoning for $I = 1$. Here the only difference is that for the $\pi\Sigma$ decay, it is c_s which is zero, so we find horizontal lines. Since we want the signs of the $I = 0$ and $I = 1$, $\pi\Sigma$ amplitudes to be opposite, as found experimentally, we shade the $+$ sectors this time.

The allowed solution must then fall in a sector where there is shading for both i-spin states, or for neither. For our example, Fig. 13c shows there is no unshaded overlap, so that only the doubly shaded overlap sectors are allowed. Fortunately the region of intersection of the lines, arrived at without sign considerations, falls in the upper overlap sector!

References

1. G. Cocconi et al., *Phys. Letters*, **8**, 134 (1964); S. L. Adelman et al., *Phys. Rev. Letters*, **13**, 555 (1964); C. M. Ankenbrandt et al., *Nuovo Cimento*, **35**, 1052 (1965); C. Bellettini et al., *Phys. Letters*, **18**, 167 (1965); E. W. Anderson et al., *Phys. Rev. Letters*, **16**, 855 (1966); I. M. Blair et al., *Phys. Rev. Letters*, **17**, 789 (1966); E. Gellert et al., *Phys. Rev. Letters*, **17**, 884 (1966); S. Almeida et al., *Nuovo Cimento*, **50A**, 1000 (1967); K. J. Foley et al., *Phys. Rev. Letters*, **19**, 397 (1967); V. Alles-Borelli et al., *Nuovo Cimento*, **47**, 232 (1967); G. Alexander et al., *Phys. Rev.* **154**, 1284 (1967); Y. Y. Lee et al., *Phys. Rev.*, **159**, 1156 (1967). See also the review by Van Hove, *Proc. 13th Intern. Conf. High Energy Physics, Berkeley, 1967*, p. 253; and A. H. Rosenfeld et al., *Rev. Mod. Phys.*, **40**, 77 (1968).
2. G. Goldhaber, *Proc. 1967 Coral Gables Conf.*, p. 190.
3. E. L. Berger, E. Gellert, G. A. Smith, E. Colton, and P. Schlein, Lawrence Radiation Laboratory Report UCRL-18110, Feb. 1968 (unpublished).
4. M. N. Focacci and G. Giacomelli, Pion Proton Elastic Scattering; CERN 66-18 (1966), unpublished.
5. J. P. Merlo and G. Valladas, *Proc. Roy. Soc. (London)*, **289A**, 489 (1966).
6. H. L. Kraybill, D. L. Stonehill, B. Deler, W. Laskar, J. P. Merlo, G. Valladas, and G. W. Tautfest, *Phys. Rev. Letters*, **16**, 863 (1966). This paper contains also a list of other references used in preparing Fig. 3. See also the references given in Table II.

7. D. Morgan, *Phys. Rev.*, **166**, 1731 (1968). (For the individual references on the data of Fig. 4, see also Tables I and II.)
8. R. D. Tripp, *Proc. Intern. School Physics "Enrico Fermi," Course XXXIII*, p. 70 (1966).
9. R. M. Sternheimer and S. J. Lindenbaum, *Phys. Rev.*, **105**, 1874 (1957); **106**, 1107 (1957); **110**, 1723 (1958); **123**, 333 (1961).
10. M. G. Olsson and G. B. Yodh, *Phys. Rev.*, **145**, 1309 (1966); G. B. Yodh and M. G. Olsson, *Phys. Rev.*, **145**, 1327 (1966).
11. R. C. Arnold and J. L. Uretsky, *Phys. Rev.*, **153**, 1443 (1967).
12. These angles could be specific as described at the beginning of this section; however, for the present application it is more convenient to define them as the polar angles Θ, Φ of the incoming beam direction in a coordinate system fixed with respect to the three final-state particles, in the overall c.m. system. Two popular choices of the axes are: (*a*) z axis along the normal to the 3-particle decay plane, x axis in the 3-particle plane (e.g., along one of the momenta $\mathbf{p}_i$, or such that it bisects the angle between $\mathbf{p}_j$ and $\mathbf{p}_k$); (*b*) z axis along one of the final state momenta $\mathbf{p}_i$, x axis perpendicular to the three-particle decay plane.
12a. Remember that the symmetry about the equator of $Y_l^m(\Theta, \Phi)$ is $(-)^{l+m}$. To see that this symmetry must be even, note that $\cos\Theta = \hat{\mathbf{n}} \cdot \hat{\mathbf{p}}$ [between normal (an axial vector) and beam (a vector)] is of course a pseudoscalar and must have zero expectation value. Hence $d\sigma/d\Omega$ must be symmetric about the equator, and every Y_l^m in its expansion must have $l + m$ even.
13. See, for example, R. D. Tripp, *Ann. Rev. Nucl. Sci.*, **15**, 325 (1965).
14. More details may be found in Ref. 7, and in L. R. Miller and P. H. Söding, Production Angular Distributions of 3-body final states, Alvarez Group Physics Note No. 652, where a table of the A_l^m in terms of the partial-wave amplitudes is given.
15. R. G. Roberts, *Ann. Physik* **44**, 325 (1967).
16. The theoretical foundation for a relativistic isobar model, using the Fadeev-Lovelace theory of composite particle scattering, has been given by D. Z. Freedman,, C. Lovelace, and J. M. Namyslowski, *Nuovo Cimento*, **43A**, 258 (1966). Relativistic formulations of the isobar model have been described and compared with experimental data, in Refs. 17–19.
17. P. G. Thurnauer, *Phys. Rev. Letters*, **14**, 985 (1965); Relativistic Analysis of Scattering Reactions with 3-Body Final States, Univ. of Rochester Report No. UR-875-119 (1966), unpublished.
18. J. M. Namyslowski, M. S. K. Razmi, and R. G. Roberts, *Phys. Rev.*, **157**, 1328 (1967).
19. B. Deler and G. Valladas, *Nuovo Cimento*, **45A**, 559 (1967); and private communication.
20. C. Lovelace, Rapporteur's talk, *Proc. 1967 Heidelberg Conf. Elementary Particles*, p. 79.
21. A. Courau, *Nuovo Cimento*, **46A**, 291 (1966).
22. B. Deler, J. P. Merlo, G. W. Tautfest, and G. Valladas, paper submitted to the 13th International Conference on High Energy Physics at Berkeley (1966), unpublished.
23. P. Bareyre, C. Bricman, and G. Villet, *Phys. Rev.*, **165**, 1730 (1968).
24. C. H. Johnson, Lawrence Radiation Laboratory Report UCRL-17683, Aug. 1967 (unpublished); H. M. Steiner, this volume.

25. W. B. Richards, C. B. Chiu, R. D. Eandi, A. C. Helmholz, R. W. Kenney, B. J. Moyer, J. A. Poirier, R. J. Cence, V. Z. Peterson, N. K. Sehgal, and V. J. Stenger, *Phys. Rev. Letters*, **16**, 1221 (1966); W. B. Richards, Lawrence Radiation Laboratory Report UCRL-16195, Nov. 1965 (unpublished).
25a. S. Sasaki, J. Takahashi, and K. Ozaki, *Progr. Theoret. Phys.*, **38**, 1326 (1967).
26. J. Anderson, F. S. Crawford, and J. Doyle, private communication; J. Doyle, Lawrence Radiation Laboratory Report UCRL-18139, 1968 (unpublished).
27. R. Armenteros, M. Ferro-Luzzi, D. W. G. Leith, R. Levi-Setti, A. Minten, R. D. Tripp, H. Filthuth, V. Hepp, E. Kluge, H. Schneider, R. Barloutaud, P. Granet, J. Meyer, and J. P. Porte, *Phys. Letters*, **24B**, 198 (1967).
28. A. Kernan and W. Smart, *Phys. Rev. Letters*, **17**, 832; 1125 (E) (1966), see also A. H. Rosenfeld, *Proc. 1966 Yalta School of Elementary Particles*, p. 153.
29. J. J. de Swart, *Rev. Mod. Phys.*, **35**, 916 (1963).
30. R. D. Tripp, D. W. G. Leith, A. Minten, R. Armenteros, M. Ferro-Luzzi, R. Levi-Setti, H. Filthuth, V. Hepp, E. Kluge, H. Schneider, R. Barloutaud, P. Granet, J. Meyer, and J. P. Porte, *Nucl. Phys.*, **B3**, 10 (1967).

Forward Dispersion Relations—Their Validity and Predictions*

S. J. LINDENBAUM

Brookhaven National Laboratory
Upton, New York

I. πN Elastic Scattering

The general treatment of πN elastic scattering and their dispersion relations are consequences of applying the special theory of relativity to the πN scattering. It has for some time been conventional to describe the πN elastic scattering by a Lorentz invariant scattering amplitude, $F(s, t)$ invariant with respect to proper and improper Lorentz transformations, such that $d\sigma/dt = |F(s, t)|^2 \cdot s, t, u$ are the well-known Lorentz invariants defined in terms of the symmetric form of the scattering diagram shown.

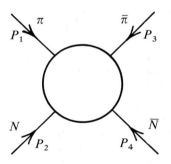

$$s = (p_1 + p_2)^2 = (p_3 + p_4)^2 = (E^p_{c.m.} + E^\pi_{c.m.})^2 = m_\pi{}^2 + m_p{}^2 + 2m_p E_{\pi \, lab} \tag{1}$$

the last formula on the extreme right being valid for the case where the proton is at rest in the lab system.

$$t = (p_1 + p_3)^2 = (p_2 + p_4)^2 = -2|P_{c.m.}|^2[1 - \cos \theta_{c.m.}] \tag{2}$$

$$u = (p_1 + p_4)^2 = (p_2 + p_3)^2 \tag{3}$$

* Work performed under the auspices of the U.S. Atomic Energy Commission.

Only two of the three quantities s, t, and u are independent, the third being determined by the relation:

$$s + t + u = 2m_p^2 + 2m_\pi^2 \tag{4}$$

In order to conserve angular momentum and parity we must construct scalars out of the three vectors at our disposal $\mathbf{p}_1$, $\mathbf{p}_2$ and the nucleon spin $\boldsymbol{\sigma}$, and we obtain the well-known result that the most general form of $F(s, t)$ involves two independent complex functions

$$F(s, t) = f(s, t) + g(s, t)\boldsymbol{\sigma} \cdot \hat{n} \tag{5}$$

where

$$\hat{n} = \frac{\mathbf{P}_1 \times \mathbf{P}_3}{|\mathbf{P}_1 \times \mathbf{P}_3|} \tag{6}$$

The $\boldsymbol{\sigma} \cdot \hat{n}$ factor of the second term of Eq. (5) contains a $\sin \theta$ factor, and $g(s, t)$ must behave regularly and remain finite near $t = 0$ for a finite force range. Hence $F(s, t \approx 0) = f(s, t \approx 0)$. There is only one nonvanishing independent complex amplitude which is a function only of (s, t).

So far, although we have required Lorentz transformation invariance, we have not explicitly applied the causality condition inherent in the special theory of relativity. It is interesting to note at this point that Einstein considered special relativity as only likely to be a correct theory for greater than molecular distances and hence it was a macroscopic relativity.

In 1946, Kronig suggested placing the requirement on the S-matrix, that signals do not travel faster than light in vacuum. It is this limitation which is generally referred to as "microscopic causality." Just as in optics some twenty years earlier, it had been found that the causality condition leads to dispersion relations connecting the real part of the forward scattering amplitude with integrals over energy of a function involving the total cross section and energy, and hence a similar result was obtained for the charged πN scattering.[1] As is well known, these early derivations did obtain the correct results; nevertheless, they were of a heuristic nature.

The crux of the matter is that if one proves (or assumes) that the scattering amplitude (for forward scattering or fixed t) is analytic* in the upper half complex energy plane, it is a straightforward matter to obtain the forward dispersion relation by a simple application of Cauchy's theorem. Similarly, given the required analyticity properties for the scattering amplitude for finite t, then a straightforward application of Cauchy's theorem leads to the fixed t dispersion relations.* It has subsequently been shown for

---

* Obviously in the case of fixed t dispersion relations, their validity is limited to the established analyticity domain. They are certainly valid within the Lehman ellipse. Recently the use of unitarity and the LSZ formalism has expanded the domain of analyticity.

an arbitrary angle that the dispersion relations can be rigorously derived from the basic principles of a local relativistic quantum field theory satisfying the following requirements*:

1. Lorentz invariance of the theory.

2. Local commutativity: The microscopic causality condition requires that if two space-time points (x_1, t_1) and (x_2, t_2) are separated by a finite spacelike distance they cannot influence one another. This causality condition is imposed by requiring that operators which correspond to physically measurable quantities shall commute for any two points separated by finite spacelike distances.

3. Mass spectrum conditions: The existence of a spectrum of positive energy stable particles is assumed, which satisfy the following relationship:

$$\text{Total energy} = |(m_\alpha{}^2 + P_\alpha{}^2)^{1/2}|$$

4. The scattering amplitude: ($F(s, t)$ can be bounded by a polynomial in s of finite power $F(s, t) <$ constant $|s|^n$ as $|s| \to \infty$ where n is a positive integer.

5. Unitarity: i.e., the S-matrix is unitary.

This assumption allows us to use the optical theorem in the forward dispersion relations. The above conditions can be met in the LSZ formalism. Recently it has been also shown that the analyticity properties for the two-particle scattering amplitudes necessary for proof of dispersion relations were also rigorously obtainable from the general set of axioms of Wightman for a local relativistic quantum field theory.[5]

Of course, if one prefers to assume the necessary analyticity properties of the S-matrix as a starting point, this obviously leads directly to the proof of the dispersion relations.

II. πN Forward Dispersion Relations

The $\pi^\pm p$ pion–nucleon forward scattering amplitudes can be expressed in terms of their real and imaginary parts as:

$$f_\pm(\omega) = D_\pm(\omega) + iA_\pm(\omega) \tag{7}$$

where ω is the total pion energy in the laboratory system where the proton is at rest, D and A are both real and $f_\pm(\omega) = kF(s, o)/\pi^{1/2}$. It is most

---

* See Refs. 2 and 3 for a more complete set of references.

convenient to employ the symmetric and antisymmetric forms:

$$D^+(\omega) = \tfrac{1}{2}[D_-(\omega) + D_+(\omega)] \qquad (8a)$$

$$A^+(\omega) = \tfrac{1}{2}[A_-(\omega) + A_+(\omega)] \qquad (8b)$$

$$D^-(\omega) = \tfrac{1}{2}[D_-(\omega) - D_+(\omega)] \qquad (8c)$$

$$A^-(\omega) = \tfrac{1}{2}[A_-(\omega) - A_+(\omega)] \qquad (8d)$$

These forms then have simple even and oddness properties by crossing symmetry when $\omega \to -\omega$. Hence using the fact that $f_\pm(\omega)$ is analytic in the entire upper half plane, applying the Cauchy theorem and assuming the integral around the infinite semicircle vanishes, one obtains:

$$D^\pm(\omega) = \frac{P}{\pi} \int_{-\infty}^{\infty} \frac{A^\pm(\omega')\, d\omega'}{(\omega' - \omega)} \qquad (9)$$

where P denotes taking the principal value of the integral. From crossing symmetry we have

$$D^\pm(\omega) = \pm D^\pm(-\omega) \qquad (10a)$$

$$A^\pm(\omega) = \mp A^\pm(-\omega) \qquad (10b)$$

which allows us to eliminate the negative frequency part of the integral.

In the nonphysical region $|\omega| \leq \mu$ for the πN case, only single nucleon states can satisfy conservation of energy and momentum, and hence requiring charge conservation, the intermediate neutron state is the only point which makes a contribution. This contribution can be accurately expressed in terms of the experimentally observable πN coupling constant using the symmetrical pseudoscalar meson theory.

The result of evaluating Eq. (9) for D^- and employing the optical theorem is:

$$D^-(\omega) = \frac{2f^2\omega}{[\omega^2 - (\mu^2/2M)^2]} + \frac{\omega}{4\pi^2} P \int_{+\mu}^{\infty} \frac{d\omega' k'(\sigma_-(\omega') - \sigma_+(\omega'))}{(\omega'^2 - \omega^2)} \qquad (11)$$

(unsubtracted forward dispersion relation)

where laboratory system quantities are employed throughout and $h = c = 1$. $\omega = $ total pion energy $ = (\mu^2 + k^2)^{1/2}$, f^2 is the renormalized unrationalized pseudovector coupling constant. This will be a valid dispersion relation provided the integral on the right-hand side converges and the integral over the infinite semicircle vanishes. These conditions will be met if

$$\sigma_-(\omega') - \sigma_+(\omega') < \text{const} \ln \omega' \qquad \text{as} \ \omega' \to \infty$$

That is, if the Pomeranchuk theorem is valid we can expect the unsubtracted D^- to be a valid dispersion relation. If the Pomeranchuk theorem is not assumed we will need an additional subtraction in D^-, and in fact the early work in dispersion relations did use a singly subtracted D^-. In the case of D^+ if the total cross sections approach constants or follow the Froissart bound $\sigma \lesssim \text{const} \ln^2 \omega$ as $\omega \to \infty$ the unsubtracted forward dispersion relation will obviously not converge and one subtraction usually taken at $\omega = \mu$, is introduced to improve the convergence. The result for the once subtracted D^+ is:

$$D^+(\omega) = D^+(\mu) + \frac{f^2 k^2}{M[1 - (\mu/2M)^2][\omega^2 - (\mu^2/2M)^2]}$$

$$+ \frac{k^2}{4\pi^2} P \int_{+\mu}^{\infty} \frac{d\omega' \omega'}{k'} \frac{\sigma_-(\omega') + \sigma_+(\omega')}{\omega'^2 - \omega^2} \qquad (12)$$

(once subtracted forward dispersion relation)

D^+ will obviously converge if $\sigma_- - \sigma_+ < \text{const } \omega^{1-\varepsilon}$ as $\sigma \to \infty$, where ε is a small but finite fraction. One should note that we have not made use of charge independence in deriving these relations.* Obviously the dispersion relations for individual charge states in terms of $D_\pm(\omega)$ can be obtained by adding and subtracting D^+ and D^-.

III. Asymptotic Bounds on the Total Cross Sections

It is obvious from the foregoing that a knowledge of asymptotic bounds on the total cross sections is essential for obtaining convergent and hence valid forward dispersion relations. The reader is referred to Refs. 2 and 3 for a detailed discussion of both the experimental evidence and theoretical arguments for various asymptotic bounds and theorems regarding the total cross sections.

For the present purposes it is sufficient to state that the Froissart bound was obtained in 1961 and used unitarity and the Mandelstam representation to obtain upper bounds for the scattering amplitudes and total cross sections at high energies. The result was:

$$\sigma_{\text{total}} \leq \text{const } (\ln \omega)^2 \qquad \text{(Froissart bound)} \qquad (13)$$

This bound has recently been established from analyticity properties proved from axiomatic field theory (LSZ) and polynomial boundedness $F(s, t) < s^n$ for $s \to \infty$.[6]

---

* A bad failure of charge independence would mean that the f^2 term can be different in the D^+ and D^- relations.

IV. Early Applications of Forward Dispersion Relations

The major previous applications of the πN forward dispersion relations were made in the low energy region of 0 to several hundred to the order of 1 BeV incident pion energy. One of the first applications was to require satisfaction of these relations to select between the six nonunique phase shift solutions obtained for low energy πN scattering.[7] However, this still left the Fermi set (which has a large phase shift in the $T = \frac{3}{2}, J = \frac{3}{2}$ state and the Yang set which has a large phase shift $T = \frac{3}{2}, J = \frac{1}{2}$ state.

Forward dispersion relations for the derivations of the spin-flip amplitude[8] were used to select the Fermi set.[9] It was subsequently pointed out[10] that there remained two other possible solutions, namely, the Minami ambiguity applied to the Fermi set and a new set obtained by applying the Minami ambiguity to the Yang set. It was also possible by requiring satisfaction of the spin-flip forward dispersion relations to rule out both of these additional sets, and thus demonstrate that the Fermi set was the only solution which satisfied the dispersion relations.[10]

Although from time to time discrepancies have appeared, it has been possible eventually to obtain a self-consistent and reasonably accurate determination of πN low energy parameters and phase shifts using πN forward dispersion relations and to some extent fixed t dispersion relations.[11,12] The latter among other difficulties suffer from the fact that a subtraction introduces an arbitrary function of momentum transfer instead of a constant as is the case for the forward dispersion relations. A knowledge of both the ordinary forward scattering amplitude as well as a knowledge of spin-flip amplitudes is required as $|t|$ becomes appreciable. The estimated forward charge exchange scattering was found to be consistent up to about 1 BeV with dispersion relation predictions assuming charge independence. Because of their successful application to the phase shift analyses and determination of the coupling constants, etc., it was generally assumed that the πp forward dispersion relations were valid at least up to a few hundred MeV of incident pion energy. There were several points of doubt one could raise as to the validity of the earlier conclusions.

1. The uncertainty of the effects of unforseen asymptotic behavior. Neither the experimental evidence for, or theoretical understanding of asymptotic behavior was in good shape at this time.

2. The forward scattering amplitude was not measured directly (except in isolated cases) in these experimental checks, and was deduced from the best fitting set, as a function of energy, of a number of non-unique phase shift analyses.

3. The low energy region was overwhelmed by the $T = J = \frac{3}{2}$ resonance and the results were sensitive to the low energy parameters which were deduced from these same data.

4. If the causality condition failed within a fundamental length corresponding to a few hundred MeV incident energy the large effect of the low energy resonances and low energy parameters would tend to mask its effects in the region where the most extensive experimental checks were made.

Therefore it became clear[13,14,2,3] that a critical test of the πN forward dispersion relations really required a direct precision measurement of the forward $\pi^{\pm} p$ scattering amplitude as a function of energy over a broad enough range of very high energies, and a precision measurement of the $\pi^{\pm} p$ total cross section to allow evaluation of the dispersion integrals, and also some demonstration of insensitivity of the conclusions to asymptotic behavior. The advantage of going to the highest energies possible are the following:

1. We can introduce additional subtractions at the highest energy to virtually eliminate sensitivity to even highly implausible and unphysical behavior of the high energy total cross sections.

2. We become very insensitive to the errors in the low energy parameters and total cross-section data.

3. As we shall see in Section VII in the discussion of the effects of failures of the microscopic causality condition (i.e., introduction of a fundamental length), the check of the validity of the πN forward dispersion relations to an incident maximum energy ω_{max} establishes that if there is a fundamental length l then

$$l < 1/\omega_{max}$$

For $-t < 0.2$ (BeV/c) several experiments[15,2,3] demonstrate that F_N can be well represented as follows:

$$F_N = \exp[a/2 + (b/2)t + (c/2)t^2] \qquad (16)$$

where the quadratic term is small compared to the linear term.

$$b \sim 8\text{--}10\,(\text{BeV}/c)^{-2} \quad \text{and} \quad c \sim 2\text{--}3\,(\text{BeV}/c)^{-4}$$

Therefore it is clear from the above formulas and the observed values of $d\sigma/dt$ ($t \approx 0$), that the known real coulomb amplitude (which is infinite at $t = 0$) and drops as $1/|t|$ will become equal to the slowly varying nuclear amplitude at $t \approx 0.003$, and the coulomb amplitude will then rapidly become unimportant at higher $|t|$. Hence, where the coulomb amplitude is comparable with the nuclear amplitude we may expect to measure by coherent interference both the value and sign of the real part of the nuclear scattering amplitude.

Since $t \approx 0.003$ corresponds to an impact parameter ≈ 3.5 fermis which is well outside the range of nuclear force we can expect by observing these interference effects at these small $|t|$ to measure the real part of the scattering amplitude at large enough distances so that a reliable extrapolation to its final value at $t = 0$ can be made. The imaginary part of the forward scattering amplitude can be measured by high precision total cross-section measurements, and then making use of the optical theorem which depends only on unitarity, we can write in the units we are employing:

$$\text{Im } F_N(t = 0) = \sigma/4(\pi)^{1/2} \tag{17}$$

Of course, observation of the differential scattering near $t \approx 0$ also determines the imaginary part of the ordinary scattering amplitude near $t = 0$ and we can use the above information to check the consistency with the $t = 0$ value determined from total cross-section measurements.

As we shall see later, the experimental data is very consistent with assumptions that at small $|t|$

$$\text{Re } F_N/\text{Im } F_N = \alpha \tag{18}$$

is a constant, and we shall assume and justify this point later. Using Eqs. (16)–(18) we can express the differential cross section as:

$$\frac{d\sigma}{dt}(\pi^{\pm}p) = \frac{F}{|t|^2} \mp \frac{2F}{|t|} \text{Im } F_{N\pm})[\alpha_{\pm} \cos 2\delta \pm \sin 2\delta]$$

$$+ [1 + \alpha_{\pm}^2][\text{Im } F_{N\pm}]^2 + \text{multiple scattering correction} \tag{19}$$

The original calculation for δ was made by Bethe[16] and gave

$$\delta = (e^2/hc\beta)(\ln 1.06/ka\theta) \tag{20}$$

Although it was originally referred to as nonrelativistic, it turned out that the result is approximately relativistically correct also.

Over the years a number of attempts [17-19] have been made to improve the Bethe formula (i.e., make it relativistic and derive it from a more fundamental relativistic quantum electrodynamic field theory[19] point of view). Where sizeable differences have appeared[17,18] they have turned out to be errors* in the calculations. Three calculations agree within 15% with the above result. In a recent comprehensive analysis of the problem from

---

* In Ref. 17 magnetic effects in the cms system were not taken into account as pointed out by Bell. In Ref. 19 when making the infrared approximation a form factor effect was neglected.[20]

the point of view of relativistic quantum field theory, Yennie and West[20] demonstrated that the form of the Bethe formula was correct. The magnitude was correct within the limitations of precision of possible higher order corrections which would introduce corrections of less than twice the above spread in values (i.e., $\Delta\delta \lesssim 0.005$).

The demonstration of the existence of sizable real parts of the $\pi^\pm p$ forward scattering amplitudes at high energies occurred in 1964 in a series of preliminary measurements.[13,14] These results did not allow a conclusive check of the forward dispersion relations due to the large systematic errors associated with the analysis to obtain the real part of the forward scattering amplitude. The accurate measurement of the real part of the forward scattering amplitude at high energies is an extremely difficult technical problem.

Since the whole coulomb interference effect occurs in a few milliradians ($\ll 1°$) near the forward direction where the steep $|t|$ dependence of the coulomb and multiple scattering and the incident beam are present, high angular and momentum resolution, high and accurately known detector efficiency and the ability to measure and analyze an enormous number of events (mostly in the coulomb region) are all required to obtain an accurate measurement of the real part of the forward scattering amplitude. An overall absolute accuracy of 1% or better was required and obtained in the new differential scattering measurements[2,21] which were planned in 1964 and run at the Brookhaven AGS by the author's group in 1965–1966. A highly accurate knowledge of the imaginary part of the forward scattering amplitude is also required in the analysis of the small angle scattering experiments, and also to calculate the dispersion relation predictions and 0.3% was obtained in the new total cross-section measurements performed.[22]

Only the on-line computer techniques originally developed and introduced by the author's group in 1962 made experiments of this precision possible. The experimental arrangement is shown in Fig. 1. About 400–500 counter hodoscopes were employed in a magnetic spectrometer system which allowed 10^{12} possible counter combinations to cover $\sim 10\%$ of the available solid angle in the $|t|$ range of interest and allowed the data handling system to record ~ 3 million trigger selected events per hour, which led to about 3 billion events being recorded in this investigation. The overall average resolutions including apparatus, beam spread, and multiple scattering were 0.4 mrad (angular) and 0.4% (momentum). Pions were selected by Čerenkov counters and all contamination including muons were essentially eliminated to a high degree and the apparatus was made virtually accident-proof. The absolute efficiency was experimentally determined to high precision and the overall scale uncertainty of 1% was conservatively deduced from long term variations observed in these

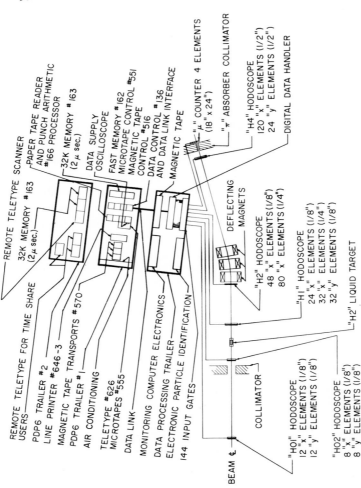

FIG. 1. A simplified sketch of the counter hodoscope on-line computer system for studying small angle scattering, utilizing a high resolution, high data rate capacity magnetic spectrometer. Hodoscopes HO_1 and HO_2 measure the location and direction of the particle before scattering in the hydrogen target. Hodoscopes H_1 and H_2 determine its location and direction after scattering and reject events not coming from the target. Hodoscope H_4 after the magnets then determines its momentum. The digital data handling equipment is in the 40' × 10' trailer near the hodoscopes. The on-line PDP-6 computer system is contained in the other 40' × 10' trailers. The apparatus is about 400' long.

efficiency measurements. Between AGS pulses the data was transferred from a buffer memory to a PDP-6 computer for on-line analysis and in parallel onto magnetic tape for a permanent record. The computer program calculated the scattering angle and momentum of each particle and constructed momentum spectra for each of 30 scattering angle bins of width 0.4–1.0 mrad and automatically accurately subtracted the frequently measured empty target backgrounds which were generally less than 10%, but became as large as 50% at the smallest angles measured. The inelastic background for hydrogen which crept under the elastic peak is the more serious background since there is some uncertainty in its subtraction. Figure 2 shows an example of the elastic peak with the inelastic background blown up by a factor of 10. Due to the high resolution the inelastic background under the elastic peak was ≲1% and its uncertainty a small fraction of this. The resolution effects were taken into account by calcu-

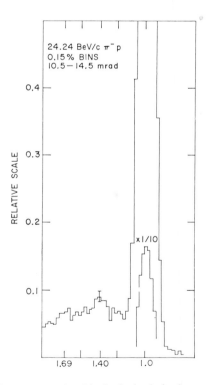

FIG. 2. Examples of the elastic peak with the inelastic background blown-up a factor of 10. The usual peak at 1.4 BeV which previously was observed only in pp is clearly prominent in $\pi^{\pm}p$ interactions throughout the s and t range covered by this experiment, thus strongly implying that it is a bonafide isobar.[23]

lating the effective values of t for the various bins by a Monte Carlo program which constructed events using the known geometry, incident particle distribution, and momentum spread.

Typical differential cross-sections are shown in Fig. 3. The errors shown are largely statistical but point to systematic errors due to the uncertainty in the determination of t and the measured nonuniformity of the hodoscopes are included. In addition there is an uncertainty in the overall absolute normalization of 1%. The multiple and plural scattering correc-

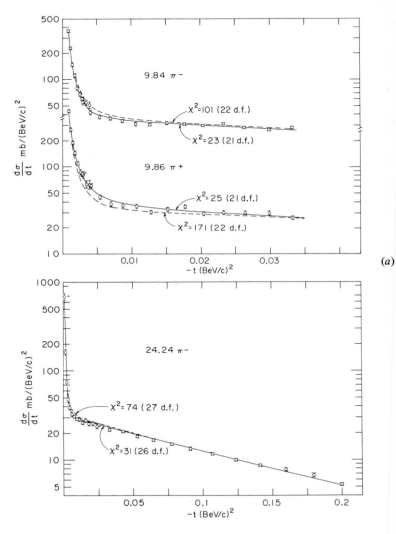

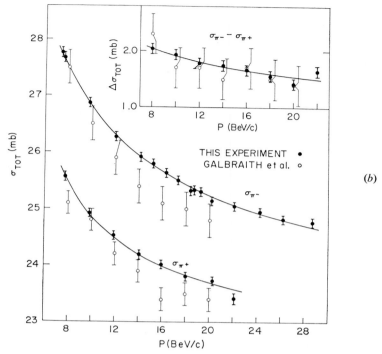

FIG. 3. (*a*) Typical differential cross sections. The 9.86π^+ data are displaced down one decade. The solid curves are best fits varying α and b. The dashed curves are best fits constraining α to be 0. The number of degrees of freedom (*d.f.*) of each fit is shown. (*b*) The new measurements of the total cross sections[22] compared to the previous data.

tions were calculated using a modified Moliere distribution. Radiative corrections[24] are small. Differences between earlier reported results[2,14] especially at low momentum and later reported results[3,21] were primarily due to second order corrections for the multiple and plural scattering terms. One of the difficulties is that scattering from the electrons is discriminated against by our excellent momentum resolution.

In Eq. (19) α and b were allowed to be free parameters. Although c was originally allowed to be a free parameter the quadratic term contributed very little in our t range and we therefore believe it to be a better procedure to fix $c = 2.4\ (\text{BeV}/c)^{-4}$ which is an average value determined from our previous experiments[15] at higher t. The values of α are very insensitive to the value of c. The values of Im $F_N(t = 0)$ were determined from precision total cross-section measurements[22] (described subsequently) and the optical theorem. Figure 3 shows examples of fits to the data. The solid curves are the best fits. The dashed curves are the best fit with $\alpha = 0$ as a constraint which is obviously ruled out by the χ^2 values. The interference is destructive for $\pi^- + p$ and constructive for $\pi^+ - p$. Hence the real part of the amplitude is always negative. The value of α is very insensitive to the minimum value of $|t|$ used in the fit. This shows that the parametrization used including the assumption of no $|t|$ dependence of α is reasonable. A minimum cutoff at $|t| = 0.001$ $(\text{BeV}/c)^2$ was selected for the final determinations. That the value of α determined then corresponds to the value of α at $t = 0$ is well justified since we have measured to $|t|$ values which correspond to impact parameters of ~ 6.5 fermis, which is well outside the nuclear force range. At the lower momenta where we are able to measure to considerably lower $|t|$, we still find no sizable effects on the values of α. To say this another way, if we wish to produce even a 5% change (which is small compared to the error) in the extrapolated value of α in going from $|t| = 0.001$ to $t = 0$, this would require that the b parameter for the real amplitude be increased by a factor of 10 compared to the b parameter for the imaginary amplitude which is obviously not to be seriously expected since this would correspond to scattering from a range of interaction of >3 fermis, and the largest nuclear force range corresponding to the lightest known nuclear particle the pion is ≈ 1 fermi. The values of b obtained in the fits are consistent with those obtained previously in larger $|t|$ experiments.[37]

The values of $\alpha_\pm$ computed according to Eq. (19) are compared in Fig. 4 to the dispersion relation predictions [using Eqs. 11)–(12)]. The

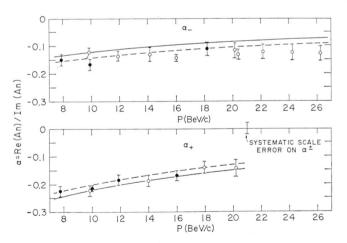

FIG. 4. $\alpha_\pm$ versus momentum. The solid curves are predictions of dispersion relations discussed in the text. The dashed curve is the result of displacing the solid curve by 0.02, the systematic scale error. The solid points represent measurements with a rearrangement of the apparatus to cover a larger angular range.

solid points represent measurements with the apparatus rearranged to cover a larger angular range with consequent worse (by $\sim 50\%$) angular resolution. As can be seen the agreement between the two setups is good demonstrating the insensitivity of the results to the resolution. The errors shown are those obtained from the least squares fit. In addition there is an overall systematic error of ± 0.02.* This error is due to uncertainty in total cross-section measurements, inelastic background subtraction, uncertainty in overall efficiency, and uncertainty in the multiple scattering corrections. This error is mostly of a scale nature and since the sign of the interference is opposite for π^+ and π^- this scale error causes opposite sign effects in α_+ and α_- and hence nearly cancels in their sum and in D^+. However, for the same reason this scale error adds in the difference of α_+ and α_- and in D^-.

Using the counter hodoscope and on-line computer system new high precision ($\sim 0.3\%$ absolute) total cross-section measurements were made from 8 to 26 BeV/c.[22] These new experimental procedures virtually eliminated the extrapolation error, and allowed accurate corrections for coulomb-nuclear interference, multiple scattering, muon contamination, etc., and allowed a virtually accident-proof apparatus. The resulting cross sections are shown in Fig. 1. The errors on the cross sections are absolute errors compounding all sources and are mostly of a systematic nature.

Table I shows the results of fitting the total cross sections to three

TABLE I

The Results of Three Least Squares Fits to the $\pi^\pm p$ Total Cross-Section Data (8–29 BeV/c) Using Power Laws of the Momentum with a Total of Five Parameters as Illustrated in the Table and Discussed in Text

Fit	$\sigma = A + B/p^C$	A	B	C
I	$\sigma_{\pi+}$	22.60 ± 0.40	25.9 ± 9.6	1.06 ± 0.24
	$\sigma_{\pi-}$	22.60 ± 0.40	19.6 ± 1.8	0.67 ± 0.08
II	$\sigma_{\pi-} + \sigma_{\pi+}$	44.24 ± 1.88	37.0 ± 8.5	0.69 ± 0.21
	$\sigma_{\pi-} - \sigma_{\pi+}$	0	3.85 ± 0.56	0.31 ± 0.06
III	σ_3	22.52 ± 0.51	24.8 ± 10.7	1.02 ± 0.28
	σ_1	22.52 ± 0.51	19.86 ± 1.92	0.58 ± 0.08

* An analysis of possible errors due to neglect of the spin-flip amplitude has been made using πN polarization, and the present and other πN differential cross-section data and it has been shown that the possible error due to neglect of the spin-flip term is quite small compared to the statistical and systematic errors on α and can be neglected.

variations of a five parameter power law fit in which all cases assume $\sigma_{\pi^-} - \sigma_{\pi^+} \to 0$ as $P_{\text{lab}} \to \infty$. Allowing an additional parameter to provide for a finite cross-section difference in the fits at infinity gives results consistent with zero difference within the errors.

VI. Comparison of Experimental Results with the Predictions of the πN Forward Dispersion Relations

Fit I was used above 8 BeV in the dispersion relation calculations shown in Figs. 4–8. Fit III would give very similar results. Fit II which is similar to a three pole P, P', ρ Regge fit* due to its very slow convergence

$$\alpha_{P'}(0) \approx 1 - 0.69 \pm 0.21 \approx 0.31 \pm 0.21$$
$$\alpha_{\rho}(0) \approx 1 - 0.31 \pm 0.06 \approx 0.69 \pm 0.06$$

of the difference gave much too large theoretical values of D^- compared to the experimentally determined values, and therefore was ruled out. However, one should note that with two standard deviations† on the exponent of the difference in Fit II, that it could be brought close to agreement with the experimental values of D^- within the systematic error on D^-.

As can be seen in Fig. 4, within the systematic scale error, we get a generally good agreement with Fit I used in the dispersion relations. Within the total cross-section errors this extrapolation can be adjusted to give an even better agreement between the prediction and the data. As previously noted, all important systematic errors including the uncertainty in δ, tend to make about equal and opposite contributions to α_- and α_+ and hence they must to a large extent cancel in the experimentally determined D^+. Furthermore, the calculated D^+ is quite insensitive to the differences between Fit I, Fit II, and Fit III, while on the other hand, D^- is extremely sensitive to systematic errors and fit differences. Hence a more critical test of the validity of the forward dispersion relation can be made by comparing the experimental results with the D^+ relation which is done in Fig. 5a, and we see that the agreement is good. If the total cross section obeys the Froissart bound (the data are certainly consistent with this) the singly subtracted D^+ dispersion integral converges reasonably well. However in order to greatly reduce the sensitivity of the integral to the high energy cross-section behavior beyond the measured momentum range, an additional subtraction was made in D^+ at 20 BeV/c and the result is shown as a dotted line in Fig. 5a which fits the data excellently.

---

* In Regge terminology.

† For these systematic errors it is our opinion that about two standard deviations represent the allowable range of variation.

This doubly subtracted D^+ dispersion relation is virtually independent of the high energy behavior of the total cross sections. To demonstrate this Fig. 6a shows the small effects of drastic and what most of us would agree, are unphysical assumed changes in the behavior of the high energy total cross sections The two assumptions made are that for $P > 35$ BeV/c (1) the total $\pi^\pm p$ cross sections both vanish; (2) both total cross sections increase faster than linearly (i.e., as $P^{1.1}$), since these totally unreasonable changes introduce only small changes* in the predicted values of D^+. It is therefore clear that we are virtually independent of any acceptable changes in the asymptotic cross-section behavior. Furthermore, variations of the low energy total cross sections and parameters within their uncertainties make very little difference in the dispersion relation predictions in the high energy range we are interested in. Hence we have critically demonstrated the validity of the doubly subtracted D^+ forward dispersion relation to beyond 20 BeV/c. This constitutes a reasonable experimental proof† of the validity of the doubly subtracted D^+ forward dispersion relation to beyond 20 BeV/c. Hence it follows that we have at least proven that $F(s, t \approx 0)$ must be analytic‡ at least for all s smaller than the value of s corresponding to 20 BeV/c incident pions. It is clear that we have probably demonstrated considerably more about the analyticity properties of scattering amplitudes than the minimum easily provable claim made above. (See, for example, the footnote at end of Section VII.)

As previously demonstrated (Fig. 5a) the singly subtracted D^+ dispersion relation is valid provided the asymptotic behavior of the high energy total cross section does not change considerably from the behavior of the good parametric fits obtained for the data. The sensitivity of the singly subtracted D^+ for asymptotic behavior of total cross sections can be deduced from Fig. 6b. Although this figure shows the sum of α_- and α_+ (instead of D^+) since these earlier calculations were available, this combina-

---

* The curvature is of no significance merely reflecting the discontinuous changes introduced in the total cross-section behavior. The additional subtraction was taken at the top of the error flag of the 20 BeV/c points. In the bottom curve the additional subtraction was taken through the center of the point.

† The author knows of no absolute experimental proofs and hence all experimental proofs can at most be classed as reasonable.

‡ It should be noted that obviously if a theory or model obtains from the data a properly analytic scattering amplitude as a functcion of s, the imaginary part of which is fit to the data, then due to the analyticity properties of the amplitude, it would give the correct real part. Therefore the fact that the Regge three pole model which uses an analytic amplitude which is adjusted to match the experimentally measured behavior of the imaginary part of the amplitude (i.e., total cross section) then gives about the right real part at high energies, is not to be taken as an additional independent success of the model.

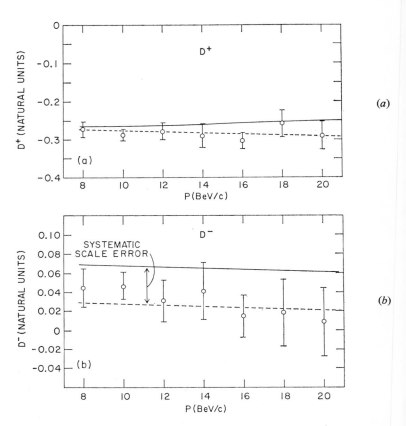

FIG. 5. (a) D^+ versus momentum. The solid curve is the prediction of the dispersion relations using the best fit to the total cross section for extrapolation. The dotted curve is the result of a subtraction at 20 BeV/c. (b) D^- versus momentum. The solid curve is the prediction of the dispersion relations as in a. The dashed curve is the result of displacing the solid curve by 0.04, the systematic scale error. cms natural units with $\hbar = c = \mu = 1$ have been used.

tion is practically all D^+ with very little $\mathbf{D}^-$ mixed in. As would be expected the sensitivity is mostly to the average level of the total cross-section average at higher energies, and we can conclude that the constant at infinity deduced in our parametric fits gives good agreement and a change of $\sim 20\%$ in its value would lead to disagreements.

Figure 5b shows the calculated results for D^- for Fit I (simple individual power law fits) compared to the data from 8 to 20 BeV/c in cms natural units. Since, as previously explained, the systematic scale errors add in D^-, their effect on the comparison is shown by the dashed

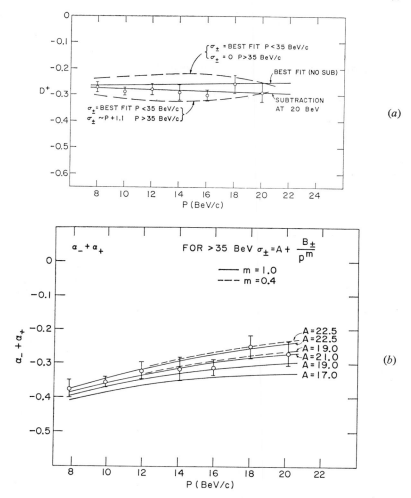

FIG. 6. (a) The D^+ dispersion relation with a second subtraction at 20 BeV/c. The small effects of even drastic and unphysical changes in the behavior of the total cross sections above 35 BeV/c are illustrated by the dashed line curves. (b) The predicted $(\alpha_- + \alpha_+)$ as a function of assumed asymptotic behavior. The experimental data is also shown.

curve* which represents a displacement of the solid curve downward by the systematic scale error which is equivalent to a change of scale of the experimental points for comparison purposes. However, as demonstrated

---

* There obviously also exists a complementary dotted line displaced upward by the systematic scale error which is of less interest since the data do not overlap it.

in Fig. 7, D^- is very sensitive to the exact power of convergence of the difference $\sigma^- = (\sigma_- - \sigma_+)/2 = \text{const}/P^n$. For example, Fit II for the total cross section which is similar to the Regge three-pole type has $n = 0.3 \pm 0.06$. As can be seen in the figure the resultant calculated D^- does not fit the D^- determined from the data. However, two to three standard derivations plus the systematic error would lead to an overlap with the data. (See Fig. 7.)

It is obvious that an effective power of convergence in the range 0.5–1.0 would generally overlap the data. Hence in spite of the appreciable scale uncertainty, the high sensitivity of the unsubtracted D^- to convergence of the cross-section difference allows us to use it as a crystal ball (assuming the dispersion relations were valid at high energies) to predict the convergence.*

VII. A Fundamental Length

The dispersion relations employed so far have been derived assuming microscopic causality holds down to infinitesimal distances and spacelike separated points cannot interfere with each other [i.e., boson (fermion) field operators at such points commute (anticommute)]. It has been shown

---

* In regard to the question of the possibility of a crossover in the total cross sections at higher energies,[25] the parametric fits I and II both allow this possibility freely. All that would be required is that the $\pi^- + p$ (or $T = \frac{1}{2}$) total cross sections fall more steeply with increasing momentum than the $\pi^+ + p$ (or $T = \frac{3}{2}$) total cross sections. Yet the fits give precisely the reverse result [i.e., $\pi^- + p$ (or $T = \frac{1}{2}$) total cross sections fall more slowly]. Of course one can never rule out unforseen changes in higher energy behavior but there is no evidence in the data that suggests or requires this. It is further clear that at low momentum the experimentally determined sign of D^- is positive even taking account of systematic errors. Furthermore, D^- seems to have the same shape as the theoretically predicted D^- but is displaced downward by about the estimated systematic scale error, Therefore it is resaonable to conclude that D^- experimental is consistent with D^- theoretical both in magnitude and sign. One should also note that it is a more fruitful procedure to plug various assumptions about asymptotic behaviour into the dispersion relations and compare with the experimental data on D^- rather than attempt to directly (or indirectly via α_- and α_+) to parametrize D^- and use and extrapolate the parametrization to higher energy since dispersion relations represent a subtle and suitable parametrization. Furthermore, an assumed parametrization

$$\alpha_+ = \sum_{n=0}^{n=n} a_n \omega^n$$

where n is a positive integer is obviously unrealistic, will diverge and may well cross over at higher energies. If one must parametrize α^+ the form

$$\alpha_\pm = \sum a_n \omega^n$$

where $n \leq 0$ including the possibility of fractions is clearly much more consonant with the form of the dispersion integrals and the behavior of the experimental data.

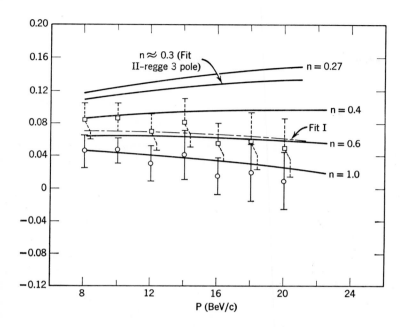

FIG. 7. The calculated values for D^- from the forward dispersion relations are shown as a function of the indicated assumed asymptotic behavior of $\sigma^- = A/P^n$ above 20 BeV/c. The open circle points represent the experimental values. The open square points represent the experimental points displaced by the systematic scale error. It should be obvious from the figure that even two standard deviations on the Regge fit $n = 0.306 \pm 0.06$ which gives $n \approx 0.42$, and maximum displacement by the systematic error on the D^- data all taken in the best direction to accommodate the Regge fit, still leads to a poor (but acceptable) fit at the high momentum.

some time ago[27] that if one relaxes this requirement and assumes an acausal region for spacelike points separated by an absolute distance $< |l|$ where $|l|$ is a so-called fundamental length, then the Cauchy dispersion integrals hold for a new defined function

$$F(\omega) = f(\omega) \exp(i|l|\omega) \qquad (21a)$$

whereas before

$$f(\omega) = D(\omega) + iA(\omega) \qquad (21b)$$

Oehme, some time ago, evaluated the result for neutral pions and it is straightforward using the crossing symmetry relation between positive and negative frequency cross sections to derive the result for charged pions.

Doing so and subtracting at $\omega = \mu$, and assuming $l \ll 1/\mu$, we obtain for D^+

$$D^+(\omega) \cos|l|\omega - A^+(\omega) \sin|l|\omega$$

$$= D^+(\mu) + \frac{f^2 k^2}{M[1 - (\mu/2M)^2][\omega^2 - (\mu^2/2M)^2]}$$

$$+ \frac{2k^2}{\pi} P \int_0^\infty \frac{\omega' \, d\omega' [A^+(\omega') \cos|l|\omega' + D^+(\omega') \sin \omega'|l|]}{(\omega'^2 - \omega^2)(\omega'^2 - \mu^2)} \tag{22}$$

When $\omega l \to 0$ these relations reduce to the ordinary forward dispersion relations. However, when ωl becomes an appreciable fraction of 1 they change drastically. The approximate calculated value of the singly subtracted D^+ for $|l| \approx 10^{-16}$ cm is shown in Fig. 8 and it is clear that even this small distance results in sizable changes in the predicted D^+ values which disagree with the experimental data values, which one should remember are almost free of systematic scale error. It is obvious that unforseen changes in the asymptotic behavior could also change the predicted D^+. However, an additional subtraction in D^+ would desensitize us to changes introduced by unforseen asymptotic behavior or a fundamental length at higher energies. However, if these were a fundamental length $|l|$ such that $\omega_l = 1/|l|$ occurred within or below our experimental observation range, due to the structure of Eq. (22) (especially the left-hand side), we would have drastic changes in both the magnitude and energy behavior of D^+ (i.e., the dispersion relations would change drastically and give completely different predictions), and therefore we can certainly conclude that if a fundamental length exists

$$|l| < 10^{-15} \text{ cm}$$

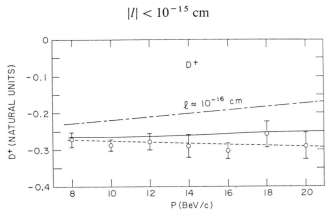

FIG. 8. The singly subtracted D^+ prediction is shown by the solid line. The dashed-dot line shows the effect of assuming a fundamental length $l \approx 10^{-16}$ cms. The points shown represent our experimental data.

There is one uncertainty in the above procedure. Due to the form of the distance between two space–time points, they can be spacelike even when separated by a large spatial distance. The form of the above changes strongly implies that the most important contributions for the acausal region come from the spacelike region which is physically close, since the changes depend on the product $\omega|l|$ indicating that high energy (i.e., accessibility to short distances) is necessary to observe the effects.

If one attempts to remove this uncertainty, one becomes model dependent. A recent particular model[28] for avoiding this uncertainty expressed the changes in the predicted values for the causal $D_\pm$ due to an acausal region characterized by $|l|$ in terms of the values of $D_\pm(\Omega)$ where $\Omega = 1/|l|$ is the lab energy corresponding to the fundamental length. Again using this procedure we can easily rule out a fundamental length occurring at energies which correspond to those within or below our range of measurement, as it would drastically change dispersion relations.

If we assume a fundamental length at higher energies and calculate its effects in the present energy range the procedure is uncertain since we must estimate by extrapolation $D_\pm(\Omega)$ as well as $\sigma_\pm(\Omega)$ and one cannot be sure that the characteristics of these (especially the former) will not change drastically at the energy which corresponds to the fundamental length. If we ignore these uncertainties and assume we can detect ~ 0.04 in D^+ we can conclude from this point of view that $|l| \lesssim 3 \times 10^{-16}$ cm.

Similar consideration can be applied to D^- but due to its extreme sensitivity to asymptotic behavior and the sensitivity of the D^- data to systematic scale errors, we believe it safer to restrict these analyses to D^+.

Hence, in summary we have been able to show directly that $|l| < 10^{-15}$ cm independent of reasonable uncertainty in asymptotic behavior. If we assume the characteristics exhibited by the parametric fits of the total cross sections data (at least for the sum) continue at higher energy we find:

$$|l| < 3 + 10^{-16} \text{ cm to } 10^{-16} \text{ cm}$$

depending on the model employed.*

VIII. Charge Independence

As previously mentioned, so far we have not made use of charge independence in either our experimental analyses or comparisons with dispersion relations. However, we can use our determination of the forward

---

* This then implies that $F(s, t \approx 0)$ is analytic for all s smaller than the value of s corresponding to 60–200 BeV/c incident pions.

scattering amplitudes and the hypothesis of charge independence to predict $d\sigma/dt_{c.e.}$ ($t = 0$). This can be compared with the estimated values of $d\sigma/dt_{c.e.}$ ($t = 0$) obtained in charge exchange experiments. Such a comparison is made in Fig. 9a. We have shown both systematic and statistical errors in our predicted values of $d\sigma/dt$ ($t = 0$)$_{c.e.}$. The estimated values from the charge exchange experiments are also shown with statistical errors only. Stirling et al.[29a] estimated a normalization error of 5–10% and Manelli et al.[29b] estimated a normalization error of ±8%.

However, perhaps the most serious uncertainty is lack of knowledge or even discussion of the uncertainties involved in extrapolating the observed $d\sigma/dt_{c.e.}$ at finite t to $t = 0$. Here as one recalls the data (see Fig. 9b) shows a rapidly t dependent shape due to the contribution of a sizeable spin-flip amplitude at small $|t|$.

In (29a) a parabolic in $|t|$ fit was used to determine $d\sigma/dt$ at $t = 0$. In (29b) the average of the first three points was taken. We have indicated the lowest $|t|$ point in (29) which is still expected to be higher than $d\sigma/dt$ ($t = 0_{c.e.}$) to show the importance of uncertainties in these estimates. An analysis of possible electromagnetic corrections should also be made.

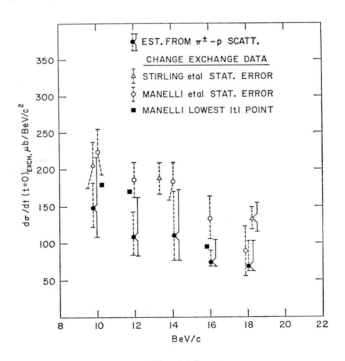

FIG. 9 (a)

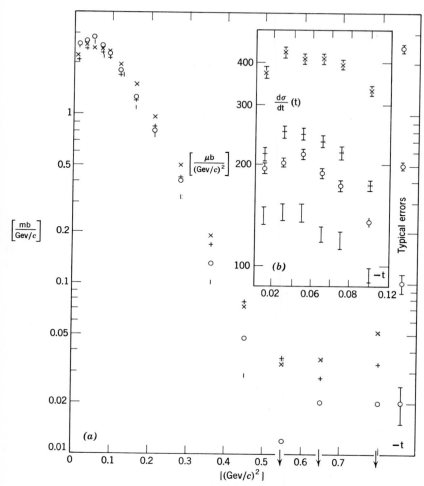

FIG. 9. (a) Assuming charge independence, the predicted values of $d\sigma/dt_{c.e.}$ $(t = 0)$ estimated from the $\pi^{\pm} - p$ small angle scattering and total cross-section measurements, are compared with the values estimated by the authors from charge exchange scattering experiments. Solid lines are systematic errors, dashed lines are statistical errors. Statistical errors are generally only shown on charge exchange points. The associated systematic errors may be larger. (b) The charge exchange data of ref. 29a. (×) 5.9 GeV/c; (+) 9.8 GeV/c; (○) 13.3 GeV/c; (|) 18.2 GeV/c.

For example, near enough to $t = 0$, the δ-phase angle correction would apply in the initial state. Nevertheless, in spite of these uncertainties we find an excellent general agreement within our stated systematic errors, and the energy behavior of the curves are reasonably similar. For the various reasons stated, I would not be surprised if the systematic errors that should be assigned to the charge exchange data should be several times as large as the statistical errors on the points. Hence, in conclusion, I believe we can be quite satisfied that we have a good verification of charge independence within the experimental errors in the range of 8–18 BeV/c. Although the uncertainties in the comparison are considerable when one remembers that the charge-exchange amplitude is $\lesssim 10\%$ the individual $T = \frac{3}{2}$ and $T = \frac{1}{2}$ nuclear amplitudes, we can at least conclude that charge-dependent terms in these individual amplitudes seem to be limited to ~ 1–2%.

IX. pp and $\bar{p}p$ Forward Scattering Amplitudes and Dispersion Relations

pp small angle scattering and total cross-section measurements were also made in the 8–26 BeV/c range. In the pp case, invariance principles limit the scattering amplitude to five independent complex functions only two of which vanish as $t \to 0$ leaving the ordinary singlet, the ordinary triplet and one spin-flip amplitude. There is a similar complexity for $\bar{p}p$. Therefore unique analysis of the experiments, and unique predictions for the dispersion relations requires the assumption of spin independence of the forward scattering amplitudes. Even with this spin independence assumption, in contrast to the $\pi^{\pm}p$ case the forward dispersion relations for the pp, and $\bar{p}p$ pair involve an extensive nonphysical region which introduces further considerable uncertainly.

Figure 10 shows the results of the new total cross-section measurements for pp.

Figure 11 shows our results[30] for α_{pp} (the ratio of the real to the imaginary part of the nuclear amplitude) and the results of previous investigations. The errors shown at each point are those obtained from the least squares fits to the data and do not include the systematic error of ± 0.02 which is mostly of a scale nature. It is clear that there is reasonable agreement with all the previous data[31] except the earlier measurements of Belletini et al.[32a] Their later remeasurement[32b] of their point at 10 BeV/c gave a lower magnitude of α in agreement with our result. If this unexplained difference were assumed to be a scale shift error in their original measurements and these were applied to their two higher points it would explain the discrepancy. When our early runs showed the above noted different behavior at high energies, additional points were measured at energies close to Belletini et al. These measurements confirmed our conclusion on the energy dependence.

The dotted line[33a] in Fig. 1 indicates the range of uncertainty in the forward dispersion relation predictions due to uncertainties in the non-

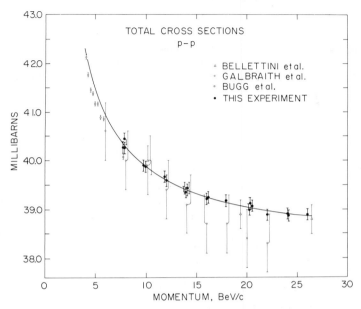

FIG. 10. High precision *pp* total cross-section measurements with a power law parametric fit.

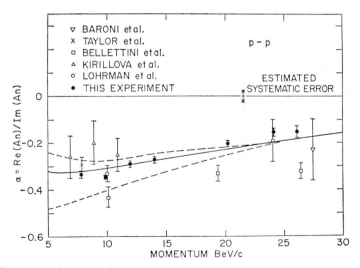

FIG. 11. α, the ratio of the real to imaginary part of the forward *pp* scattering amplitude (assuming spin independence) compared to other data, and forward dispersion relation calculations.

physical region contribution. An additional subtraction has been made at about 25 BeV. The solid line is the prediction[33b] calculated making definite assumptions (the usual poles) for the nonphysical region.

Our data certainly fits the predictions rather well. How significant this is, is not too clear in view of the spin independence assumption and the uncertainty for the nonphysical region contribution in the forward dispersion relation calculations. Nevertheless, it is quite comforting and at least implies that the pp forward amplitudes are to a large degree spin independent, and that in spite of the extensive nonphysical region, the high energy forward dispersion relation predictions are reasonable.

In regard to the spin independence assumption one should note that this gave a good fit to the experimental data. We were unable to obtain a good fit to the data with the assumption of purely imaginary amplitudes with different magnitudes and exponential slopes for ordinary singlet and triplet amplitudes. If we allowed a real part but simply relaxed the constraint that the magnitude of the singlet and triplet amplitudes be equal, we find the best fit when the amplitudes are equal, but the data are consistent with a difference as large as $\sim 30\%$ at which point α was larger by ~ 0.02.

For $\bar{p}p$ at 11.9 BeV/c we find $\alpha = -0.006 \pm 0.034$ with an additional systematic error of ± 0.06 primarily due to uncertainty in the $\bar{p}p$ total cross sections. This is in agreement within the error with the dispersion relation prediction[33b] of -0.06. Hence the forward dispersion relation prediction that $\alpha_{\bar{p}p}^-$ is quite small compared to α_{pp} is well verified.

X. "Asymptopia"

The search for a high enough energy where at least the gross features of strong interactions would exhibit an asymptotic solution (assuming there is one) has been vigorously pursued both experimentally and theoretically for some time (see Refs. 2, 3, and 13).

A few years ago I coined[34] the phrase "Asymptopia"—the theoretically promised land where all asymptotic theorems come true. One almost universal characteristic of asymptotic theories is concern with the behavior of the total cross section of a particle incident on a target nucleon, relative to an antiparticle incident on a target nucleon, for example, the pairs $(\pi^{\pm}p)$, $(pp$ and $\bar{p}p)$ and $(k^{\pm}p)$. The Pomeranchuk theorem and how it is approached is almost a universally important feature of asymptotic theories. Originally Pomeranchuk expected this theorem to be verified at ~ 10 BeV. Experiments[35,36] at the AGS and CERN soon showed this was not the case.[35]

Several years ago I became discouraged with previous attempts (for the most part of an ad hoc nature)* to determine asymptotic behavior, and concluded that we did have one precisely quantitative crystal ball that looked all the way to infinity—namely the measurements of high energy $\pi^\pm p$ forward scattering amplitudes combined with the forward dispersion relations. We have just critically demonstrated the validity of the $\pi^\pm p$ forward dispersion relations to beyond 20 BeV/c virtually independent of asymptotic behavior and demonstrated the excellent agreement between the measured real amplitudes and the singly subtracted D^+, and unsubtracted D^- in which reasonable parametric power law fits to the total cross-section measurements have been made. Therefore let us now see what we can conclude about asymptotic behavior.

Figure 12 shows the predicted asymptotic behavior of $\pi^\pm p$ total cross section for the two parametric fits (I and II) versus a 1/P scale. I

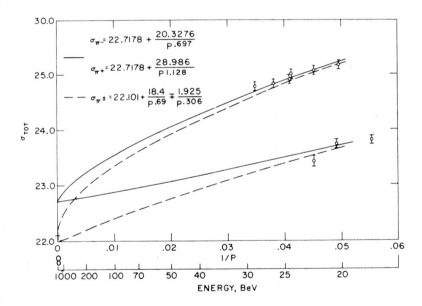

FIG. 12. The predicted values of $\sigma_{\text{tot}}(\pi^\pm p)$ above 20 BeV/c plotted versus a linear plot of 1/P in the lab system with a nonlinear energy scale (BeV) also shown. The solid lines represent a fit of type I. The dashed lines represent a fit of type II. The slight differences in parameters from Table I are due to a somewhat earlier fit and are not significant for present purposes.

* See Refs. 2, 3, 13, and 34 for further discussion.

would consider a reasonable operational definition of where Asymptopia lies as follows. The best precision we can expect even in the future to obtain in measuring the differences of π^+p and π^-p total cross sections is about 0.1%. One clear-cut characteristic of Asymptopia as mentioned before is the Pomeranchuk theorem. Therefore I will estimate where we can expect Asymptopia lies using the criterion that the Pomeranchuk theorem will be verified to best experimental precision $\sim 0.1\%$. As shown in Fig. 12 for Fit I which agrees with D^- predictions, we find this occurs at around 25,000 BeV. For Fit II (the three-pole Regge type) which disagrees badly with D^- predictions and was not used for this reason, we find this occurs at 1,600,000 BeV.

Figure 13 shows a fit similar to type I to the pp and $\bar{p}p$ total cross-sections which as you recall, fits the forward dispersion relation prediction well, assuming spin independence. Here the estimate of where Asymptopia lies is $\geq 35{,}000$ GeV.

Hence I think it safe to conclude that $\gtrsim 20{,}000$ GeV is strongly selected as determined by our crystal ball (based on the most fundamental principles) for the border of Asymptopia, as defined above.

I now believe we do have a crystal ball that can view the road to Asymptopia, and we may be able to draw relevant conclusions about it before arriving there. Our crystal ball also tells us that if microscopic causality fails at small distances (i.e., an acausal fundamental length

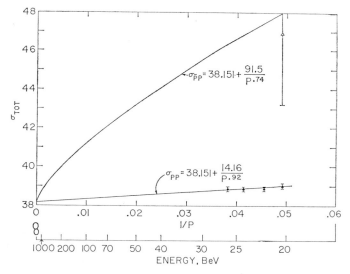

FIG. 13. The predicted values of $\sigma_{\rm tot}(pp)$ and $\sigma_{\rm tot}(\bar{p}p)$ by a fit of type I versus a linear plot of $1/p$ in the lab system with a nonlinear energy scale (BeV) also shown.

exists) that distance is definitely less than $<10^{-15}$ cm and quite likely $\lesssim 10^{-16}$ cm.

So I expect it will be a long time before we verify the Pomeranchuk theorem directly. We can use our dispersion relation calculation to predict α_- and α_+ to be observed at the higher energy accelerators at Serpukhov and Weston. Figure 14 shows these predictions. It will be extremely interesting to see if the predictions based on the present AGS experiments and the forward dispersion relations are borne out by future higher energy accelerators. I personally believe they will be.

These same experimental techniques and even the same apparatus can be used at the higher energy accelerators to check these predictions, and extend the sensitivity to detection of a fundamental length to $\sim 10^{-17}$ cm.

It is interesting to note that our present estimated limit of $\sim 10^{15}$ to 10^{-16} cm is $\sim 1/100$ to $1/1,000$ the proton interaction radius and $\sim \frac{1}{2}$ to $\frac{1}{20}$ the compton wavelength of a quark of 10 BeV mass.

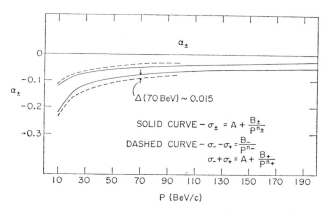

FIG. 14. The predicted values of $\alpha_\pm$ from the forward dispersion relation using Fit I (solid curve) and Fit II (dashed curve). Fit I agrees very well with the present real amplitude measurements, but Fit II (3-pole Regge type) does not.

References

1. M. L. Goldberger, *Phys. Rev.*, **99**, 979 (1955); M. L. Goldberger, H. Miyazawa and R. Oehme, *Phys. Rev.*, **99**, 986 (1955).
2. S. J. Lindenbaum, Plenary Session Review Lecture, *Proc. Oxford Intern. Conf. Elementary Particles, September, 1965.*
3. S. J. Lindenbaum, Invited Paper, *Proc. Coral Gables Conf. Symmetry Principles High Energies, January 25–27, 1967.* Pages 122–164 published by W. H. Freeman and BNL Report 11175, February 24, 1967.

4. N. N. Bogolubov and V. S. Vladmirov, *Izv. Akad. Nauk USSR Ser. Mat.*, **22**, 15 (1958).
5. K. Hepp, *Helv. Phys. Acta*, **37**, 639 (1964).
6. A. Martin, *Proc. Coral Gables Conf. Symmetry Principles High Energies, January 25–27, 1967*.
7. H. L. Anderson, W. C. Davidson, and U. E. Kruse, *Phys. Rev.*, **100**, 339 (1955).
8. R. Oehme, *Phys. Rev.*, **100**, 1503 (1955); A. Salam, *Nuovo Cimento*, **3**, 424 (1956).
9. W. C. Davidson and M. L. Goldberger, *Phys. Rev.*, **104**, 1119 (1956).
10. S. J. Lindenbaum and R. M. Sternheimer, *Phys. Rev.*, **110**, 1174 (1958).
11. J. Hamilton and W. S. Woolcock, *Rev. Mod. Phys.*, **35**, 737 (1963).
12. See related papers by Lovelace, Shaw, etc., this volume.
13. S. J. Lindenbaum, Rapporteurs Report *Proc. Intern. Conf. High Energy Physics, Dubna, 1964*, p. 188 (published by Atomizdat Moscow (1966)).
14. K. J. Foley, R. S. Gilmore, R. S. Jones, S. J. Lindenbaum, W. A. Love, S. Ozaki, E. H. Willen, R. Yamada, and L. C. L. Yuan, *Proc. Intern. Conf. High Energy Physics Dubna, 1964*, and *Phys. Rev. Letters*, **14**, 74 (1965); **14**, 862 (1965).
15. K. J. Foley, S. J. Lindenbaum, W. A. Love, S. Ozaki, J. J. Russell, and L. C. L. Yuan, *Phys. Rev. Letters*, **10**, 376 (1963); **11**, 425 (1963); **11**, 503 (1963); **15**, 45 (1965).
16. H. A. Bethe, *Ann. Phys. (N.Y.)*, **3**, 190 (1958).
17. W. B. Rolnick and R. M. Thaler, *Phys. Rev. Letters*, **15**, 177 (1965), subsequently corrected by erratum.
18. J. Rix and R. M. Thaler, *Phys. Rev.*, **152**, 1357 (1966); M. M. Islam, Report NYO-2262TA-149 and private communications; R. Serber, private communications.
19. L. D. Solov'ev, *Zh. Eksperim. Teor. Fiz.* **49**, 292 (1965) *Soviet Phys. JETP* **22**, 205 (1966).
20. D. Yennie and G. West, private communication. The author wishes to express his appreciation to Yennie and West for kindly agreeing to perform this investigation. Details will be published by these authors shortly.
21. K. J. Foley, R. S. Jones, S. J. Lindenbaum, W. A. Love, S. Ozaki, E. D. Platner, C. A. Quarles, and E. H. Willen, *Phys. Rev. Letters*, **19**, 193 (1967).
22. K. J. Foley, R. S. Jones, S. J. Lindenbaum, W. A. Love, S. Ozaki, E. D. Platner, C. A. Quarles, and E. H. Willen, *Phys. Rev. Letters*, **19**, 330 (1967).
23. K. J. Foley, R. S. Jones, S. J. Lindenbaum, W. A. Love, S. Ozaki, E. D. Platner, C. A. Quarles, and E. H. Willen, *Phys. Rev. Letters*, **19**, 397 (1967).
24. The prescription for calculating these corrections was supplied by Yennie and West (private communication). A detailed discussion of these effects will be treated in forthcoming articles.
25. T. J. Gadjdicar and J. W. Moffat, *Phys. Letters*, **25B**, 608 (1967).
26. For a review of the status of this theorem see R. Eden, *Phys. Rev. Letters*, **16**, 39 (1966).
27. R. Oehme, *Phys. Rev.*, **100**, 1503 (1955).
28. D. I. Blokhintsev, *Soviet Phys., USPEKHI*, **9**, 405 (1966).
29a. A. V. Stirling, P. Sondereggar, J. Kirz, P. Falk Variant, O. Guisan, C. Brunetson, P. Borgeoud, M. Yvert, J. P. Guillard, C. Caverzasio, and B. Amblard, *Phys. Rev. Letters*, **14**, 763 (1965).
29b. I. Manelli, A. Bigi, R. Carrara, M. Wahlig, and L. Sodickson, *Phys. Rev. Letters*, **14**, 408 (1965).

30. K. J. Foley, R. S. Jones, S. J. Lindenbaum. W. A. Love, S. Ozaki, E. D. Platner, C. A. Quarles, and E. H. Willen, *Phys. Rev. Letters*, **19**, 857 (1967).

31. E. Lorhmann, H. Meyer, and H. Winzeler, *Phys. Letters*, **13**, 78 (1964); G. Baroni, A. Manfredini, and V. Rossi, *Nuovo Cimento*, **38**, 95 (1965); A. E. Taylor, A. Ashmore, W. S. Chapman, D. F. Falla, W. H. Range, D. B. Scott, A. Astbury, F. Capicci, and T. G. Walker, *Phys. Letters*, **14**, 54 (1965); L. Kirillova, L. Khristov, V. Nikitin, M. Shafranova, L. Strunov, V. Sviridov, Z. Korbel, L. Rob, P. Markov, Kh. Tchernev, T. Todorov, and A. Zlateva, *Phys. Letters*, **13**, 93 (1964); K. J. Foley, R. S. Gilmore, R. S. Jones, S. J. Lindenbaum, W. A. Love, S. Ozaki, E. H. Willen, R. Yamada, and L. C. L. Yuan, *Phys. Rev. Letters*, **14**,. 74 (1965).

32a. G. Belletini, G. Cocconi, A. N. Diddens, E. Lillethun, J. Pahl, J. P. Scanlon, J. Walters, A. M. Wetherell, and P. Zanella, *Phys. Letters*, **14**, 164 (1965).

32b. Ibid., *Phys. Letters*, **19**, 705 (1966).

33a. I. I. Levintov and G. M. Adelson-Volsky, *Phys. Letters*, **13**, 185 (1964).

33b. P. Söding, *Phys. Letters*, **8**, 285 (1964).

34. S. J. Lindenbaum, *Proc. 2nd Coral Gables Conf. Symmetry Principles, January 20–22, 1965*, published by Freeman and Company.

35. S. J. Lindenbaum, W. A. Love, J. A. Nieberer, S. Ozaki, J. J. Russell, and L. C. L. Yuan, *Phys. Rev. Letters*, **7**, 185 (1961); **1**, 362 (1961).

36. G. Van Dardel, D. Dekkers, R. Mermod, M. Vivargent, C. Weber, and K. Winder, *Phys. Rev. Letters*, **8**, 173 (1962).

37. K. J. Foley, S. J. Lindenbaum, W. A. Love, S. Ozaki, J. J. Russell, and L. C. L. Yuan, *Phys. Rev. Letters*, **11**, 425 (1963); K. J. Foley, R. S. Gilmore, S. J. Lindenbaum, W. A. Love, S. Ozaki, E. H. Willen, R. Yamada, and L. C. L. Yuan, *Phys. Rev. Letters*, **15**, 45 (1965).

Partial Wave Dispersion Relations and N/D Calculations*

GORDON L. SHAW

University of California
Irvine, California

Recent phase shift analyses[1-5] of πN scattering have disclosed many interesting features in the low partial waves. Writing the S-matrix elements

$$S \equiv \eta e^{2i\delta}$$

we note, e.g., that for center of mass energy $W \lesssim 1700$ MeV the S_{11}, S_{31}, P_{11}, D_{13}, and F_{13} waves have been found to have large δ's and small η's (i.e., large inelastic production cross sections) in contrast to the familiar P_{33} resonance which shows only small inelastic production. As we have just heard from Steiner[6] and Lovelace[7] there may be many more "resonances"[8] at somewhat higher energies with possibly more than one in a given partial wave!

We will discuss the dynamical attempts to understand these low energy phenomena using partial waves dispersion relations. As we shall see, we cannot calculate the detailed behavior of the large δ's in terms of known πN forces. Either our knowledge of the forces is very poor or higher mass inelastic channels must be explicitly considered. There are good indications that the latter is the crucial feature. Let us start with some introductory remarks which are well-known to the theorists.

A dispersion relation for the partial wave amplitude

$$f_J{}^I(W) = [S_J{}^I(W) - 1]/2i\rho(W) \tag{1}$$

where ρ is a kinematical factor can be written as

$$f_J{}^I(W) = B_J{}^I(W) + \text{NP} + \frac{1}{\pi}\int_{\text{PC}} \frac{\text{Im} f_J{}^I(W')\,dW'}{W' - W - i\varepsilon} \tag{2}$$

B is the generalized potential obtained from the crossed $t(\pi\pi \to N\overline{N})$

* Supported in part by the National Science Foundation.

and $u(\pi N \to \pi N)$ channels containing the discontinuities across the unphysical cuts (UPC)

$$B = \frac{1}{2\pi i} \int_{\text{UPC}} \frac{\text{disc} f(W') \, dW'}{W' - W} \tag{3}$$

NP denotes the nucleon pole term which is present in the $I = \frac{1}{2}$, $J = \frac{1}{2}$ partial wave. Working in the W plane rather than the (energy squared) s plane removes $s^{1/2}$ singularities and gives both parity states in terms of a single function since (for $\rho = k$)

$$f^I_{J=(l+1)-1/2}(W) = -f^I_{J=l+1/2}(-W)$$

The physical cuts (PC) extend along the real axis for $|W| > W^T = m + 1$ (in pion units) and in this region unitarity gives the relation

$$\text{Im} f = \rho |f|^2 + (1 - \eta^2)/4\rho, \quad |W| > W^T \tag{4}$$

so that

$$f(W) = B(W) + \text{NP} + \frac{1}{\pi} \int_{\text{PC}} \frac{(1 - \eta^2(W')) \, dW'}{(W' - W - i\varepsilon) 4\rho(W')}$$

$$+ \frac{1}{\pi} \int_{\text{PC}} \frac{\rho(W') |f(W')|^2 \, dW'}{W' - W - i\varepsilon} \tag{5}$$

Ignoring inelastic effects for the moment (i.e., set $\eta(W) = 1$), we have a nonlinear integral equation to solve for f in terms of the potential B. Chew and Mandelstam introduced the N/D equations which convert (5) to a form which involves solving a linear integral equation for N and a principal value integral for Re D:

$$N(W) = B(W) + \frac{1}{\pi} \int_{-\infty}^{\infty} \frac{B(W') - (W/W')B(W)}{W' - W} \rho(W')\theta(|W'| - W^T) \, dW'$$

$$D(W) = 1 - \frac{W}{\pi} P \int_{-\infty}^{\infty} \frac{\rho(W')\theta(|W'| - W^T)N(W') \, dW'}{(W' - W)W'}$$

$$- i\rho(W)\theta(|W| - W^T)N(W) \tag{6}$$

(D has been normalized to 1 at $W = 0$. The solution $f = N/D$ is independent of this subtraction point). Straightforward, minor modifications of (6) can be made to take into account inelastic effects: (i) If we can approximate the important inelastic channels in a given partial wave by a small number n of two-body (e.g., ηN) or quasi-two-body channels (e.g., ρN, πN_{33}^*), then the multichannel $f = ND^{-1}$ equations are also given by (6) with the relevant quantities being $n \times n$ matrices (B and ND^{-1} are symmetric, ρ and θ are diagonal with ρ_i and W_i^T the kinematical factor and threshold energy for the ith channel). Thus given the potentials B_{ij} we can solve the multichannel equations (6) for all the scattering amplitudes f_{ij}; (ii) In a less ambitious calculation, if in addition to the πN potential B, the inelastic factor $\eta(W)$ is given (taken, e.g., from the phase shift analyses), one can solve the one channel N/D equations (of Froissart[9] or) of Frye-Warnock[10]:

$$\frac{2\eta(W)}{1 + \eta(W)} \operatorname{Re} N(W) = \bar{B}(W) + \frac{1}{\pi} \int_{PC} \left[\bar{B}(W') - \frac{W}{W'} \bar{B}(W) \right]$$

$$\times \frac{2\rho(W')}{1 + \eta(W')} \frac{\operatorname{Re} N(W')\, dW'}{W' - W}$$

$$D(W) = 1 - \frac{W}{\pi} \int_{PC} \frac{2\rho(W')}{1 + \eta(W')} \frac{\operatorname{Re} N(W')\, dW'}{W'(W' - W - i\varepsilon)}$$

$$\bar{B}(W) = B(W) + \frac{1}{\pi} P \int_{PC} \frac{1 - \eta(W')}{2\rho(W')} \frac{dW'}{W' - W}$$

$$\operatorname{Im} N(W) = \frac{1 - \eta(W)}{2\rho(W)} \operatorname{Re} D(W), \qquad \text{for} \quad |W| > W^T \quad (7)$$

Eqs. (6) and (7) are readily solved on a computer by standard methods.[11,12] Thus it is not necessary or desirable, in general, to use various approximations to these equations.

There are a number of technical points to consider, e.g.: (i) What should you take for ρ in order to ensure the correct behavior $\rho \operatorname{Re} f \propto k^{2l+1}$ as $k^2 \to 0$ (Note that although $\rho \operatorname{Im} f$ should vanish like k^{4l+2} at $|W| = m + 1$, it should vanish like k^2 at the conjugate point $|W| = m - 1$[13,10]); (ii) What is the correct behavior of f at $W = 0$?; (iii) The usual approximations for B diverge at high energies and some sort of cutoff in ρ, or cutting off the dispersion integral (3) for B along the unphysical cuts[14]; (iv) In the light of the above problems, is it all right to work in the s plane? However, the general remarks we will make are not dependent on these details.

One last, but crucially important, introductory remark is related to the question of CDD[15] solutions: Given B and (and NP), is the solution (7) the only solution of (5)? No! There are many possible $f(W)$ having the same $B(W)$ and $\eta(W)$ (and NP) which satisfy (5) and differ by zeroes (pairs of zeroes if at complex W) of the S-matrix on the physical sheet. These additional solutions of (5) are called CDD solutions. Note that the pole term NP does not explicitly appear in (7) (or (6)). Hopefully, it will occur as a dynamical bound state in the P_{11} wave, i.e., a zero in D (or a zero in the determinant of D for the multichannel problem (6)) at $W = m$ with the correct residue (proportional to g^2, the πN coupling constant). One can, of course, treat the nucleon as an elementary particle or CCD effect by including NP in the potential B (provided that $D(m) \neq 0$).[12]

We might also treat the resonances that one found in the phase shift analyses as CDD effects. This approach is completely unsatisfactory since we believe that the baryons and certainly the resonances are the result of the dynamics. However, as we shall discuss, the "important" inelastic channels must be *explicitly* considered in the ND^{-1} equations (6) in order to construct a dynamical model of the resonances.

It is easy to demonstrate that the one-channel equations (7) are not always equivalent to the multichannel equations (6).[16] Consider a two-channel problem with a higher mass channel 2 coupled to the πN channel 1. Suppose there were a weak diagonal πN potential B_{11} and a strong B_{22} so that with the interchannel coupling $B_{12} = 0$, a bound state appeared in f_{22} in the energy region between the two thresholds whereas f_{11} would be small. Now gradually switch on B_{12} and then in the solution of (6), an arbitrarily narrow resonance (for sufficiently weak B_{12}) will occur in the πN elastic amplitude f_{11}. Take the (weak) B_{11} potential and η (as determined from $|f_{12}|^2$) and calculate the πN amplitude f from the one-channel equations (7). The sharp resonance (or quasi-bound state) will not appear in f. We say that in order to make f and f_{11} equivalent, CDD effects must be added to f or that this narrow resonance is a CDD resonance with respect to the πN channel. (There are zeroes in the S-matrix element S_{11} on the physical sheet, not present for this example in S determined from (7) which retreat through the inelastic cut as B_{22} is made sufficiently weak, at which point the two calculations agree.) Unfortunately, there is no simple, general criterion to tell us if we must explicitly consider an inelastic channel in the multichannel formalism. However, we can make model dependent judgments in specific situations.

We note that we can use the dispersion relation (5) together with $f(W)$ as determined from the phase shift analyses to see if our approximate forms for the πN generalized potential B are consistent with the physical problem.[17] However, even if B and f satisfy (5) this does not mean we understood the dynamics. For example, with B given by a model and η taken from experiment (and joined smoothly to an asymptotic form) we then solve the N/D equations (7) for f. If our calculated f does not agree with experiment then (i) B is inadequate (at present, we can check it using (5) over a fairly limited energy range compared to the region we must know B in order to solve (7)), and/or (ii) CDD effects are present, i.e., higher mass inelastic channels must be explicitly considered.

Now let us turn to the calculations. Define partial wave projections of the invariant amplitudes A and B:

$$[A_l(s), B_l(s)] = \int_{-1}^{1} d\cos\theta [A(s, t), B(s, t)] P_l(\cos\theta) \qquad (8)$$

and use the superscript L to denote that of the amplitude which is regular in the physical s region. The πN generalized potential $B_J^I(W)$ is then given by

$$B_{J=l\pm 1/2}^I(W) = \frac{1}{\rho(W)} \frac{k}{16\pi W} \{(E + m)[^L A_l^I(s) + (W - m)^L B_l^I(s)]$$

$$+ (E - m)[-^L A_{l\pm 1}^I(s) + (W + m)\, ^L B_{l\pm 1}^I(s)]\} \qquad (9)$$

where E is the energy of the nucleon.

It has been usually assumed[14,18,19] that the N and $N_{33}{}^*$ (1236 MeV) exchange dominate the u channel forces (Fig. 1a, b) and the exchange of the ρ and the S-wave, $I = 0$, $\pi\pi \equiv$ "σ" dominate the t channel forces (Fig. 1c, d). In the narrow width approximation, the exchange of N, $N_{33}{}^*$, ρ and σ yield the following contributions to the invariant amplitudes A

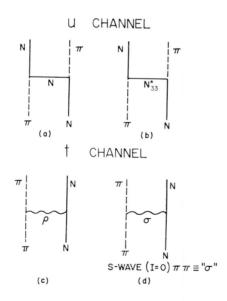

FIG. 1. Forces in πN scattering.

and B

$$^{L}A^{(I=\frac{1}{2},\, I=\frac{3}{2})}(s,\, t) = 0 \tag{10a}$$

$$= \left(\frac{4}{3}, \frac{1}{3}\right) \frac{8\pi\gamma_{33}}{m_{33}^2 - u} \left\{ (m_{33} - m)(E_{33} + m)^2 \right.$$

$$+ \frac{3}{2}(m_{33} + m)$$

$$\left. \times \left[m^2 + 1 - \frac{1}{2} m_{33}^2 - s + \frac{(m^2 - 1)^2}{2m_{33}^2} \right] \right\} \tag{10b}$$

$$= \left(1, -\frac{1}{2}\right) \frac{12\pi\gamma_2}{m_\rho^2 - t} (2s + t - 2m^2 - 2) \tag{10c}$$

$$= \left(\frac{1}{\sqrt{6}}, \frac{1}{\sqrt{6}}\right) \frac{g_{\sigma\pi\pi} g_{NN\sigma}}{m_\sigma^2 - t}, \tag{10d}$$

$$^{L}B^{(I=\frac{1}{2},\, I=\frac{3}{2})}(s,\, t) = \left(\frac{1}{3}, -\frac{2}{3}\right) \frac{3g^2}{m^2 - u} \tag{11a}$$

$$= \left(-\frac{4}{3}, -\frac{1}{3}\right) \frac{8\pi\gamma_{33}}{m^2 - u} \left[-(E_{33} + m)^2 \right.$$

$$\left. + \frac{3}{2}\left(m^2 + 1 - \frac{1}{2} m_{33}^2 - s + \frac{m^2 - 1}{2m_{33}} \right) \right] \tag{11b}$$

$$= \left(1, -\frac{1}{2}\right) \frac{-24\pi(\gamma_1 + 2m\gamma_2)}{m_\rho^2 - t} \tag{11c}$$

$$= 0 \tag{11d}$$

where m_{33}, m_ρ, and m_σ are the masses of the N_{33}^*, ρ and σ, respectively, and E_{33} is the nucleon energy at the position of the resonance; $g^2/4\pi = 14.6$ and the N_{33}^* residue $\gamma_{33} = 0.06$ corresponds to a width of ~ 120 MeV. The residues γ_1 and γ_2 proportional to the electric and magnetic couplings of the ρ to the nucleon, as determined from a fit to the nucleon's isovector electromagnetic form factor, are $\gamma_2 \approx 0.27\gamma_1$ and $\gamma_1 \sim -1.0$.[20,21] The parameters of the σ are the only ones not well known. Indeed, the pole approximation (10d) is probably not a very accurate representation of the $I = 0$ $\pi\pi \to N\bar{N}$ amplitude.[7] However, reasonable estimates indicate that the contribution to the potential from σ exchange is not very important for the low partial waves.[17] It should be quite important for the high J states. (We will return to this point later.)

Consider the πN generalized potential $B_J{}^I$ to be given by N, $N_{33}{}^*$ and ρ^{22} exchange as calculated from Eqs. (9)–(11). We then use this B as input into the one channel N/D equations. These equations are then solved numerically to obtain the phase shifts $\delta_J{}^I$. As mentioned previously, there are a number of practical points to consider. We present the results of some calculations[12] which (i) are done in the W plane (recall that $-W$ region yields the opposite parity state); (ii) take the phase space, factor

$$\rho = (E + m)(k^2/s)^J$$

and (iii) handle the high energy divergence of $B(W)$ by replacing the upper limit of ∞ in the integral over the physical cut by a finite value Λ. Thus we have one parameter, Λ, which is adjusted to give the best fit for a given partial wave I, J. The cutoffs Λ were found to occur at moderately high energy and thus can compensate for a potential which differs at low energy from the physical situation by a slowly varying function of energy. The quantitative results are of course sensitive to the value of Λ, but we observe that the Λ's did not differ greatly for the different J states. Reasonable fits as shown in Figs. 2 and 3 are obtained to the (low J) phase shifts at energies when the δ's are relatively small. It would appear that these one-parameter fits to the pairs $(l = J \pm \tfrac{1}{2})$ of phase shifts are nontrivial and have physical significance.

The most striking theoretical results for πN scattering come from the Chew-Low[23] reciprocal bootstrap which take N and $N_{33}{}^*$ exchange as the input forces producing the N as a P_{11} bound state as well as producing the P_{33} resonance. In the static limit and taking the D functions to be linear, they predict a ratio of g^2 divided by γ_{33} which is in beautiful agreement with experiment. Unfortunately, the full N/D calculations for the 11 and 33 waves do not give good agreement: In each case, forcing the N and $N_{33}{}^*$ by adjusting the cutoff, the residues are about 60 to 100% too large. (It is interesting to note that Regge slopes for the N and $N_{33}{}^*$ determined by doing the calculations for noninteger J are too small by the same factor.) Further, in Figs. 4 and 5 we see that the P_{33} phase shift looks nothing like experiment above the resonance and the P_{11} phase shift shows no sign of the well-established Roper resonance. It is amusing that for the 11 wave, good results can be obtained by inserting the N as an "elementary particle" or CDD effect.[24]

The one-channel calculations at present are not in good agreement with any of the large δ's. A good possibility is that the input B's discussed above are inadequate. For the P_{11} wave, the discrepancy,

$$\Delta(W) \equiv \operatorname{Re} f(W) - B(W) - \mathrm{NP}(W) - \frac{P}{\pi} \int_{\mathrm{PC}} \frac{\operatorname{Im} f(W')\, dW'}{W' - W} \qquad (12)$$

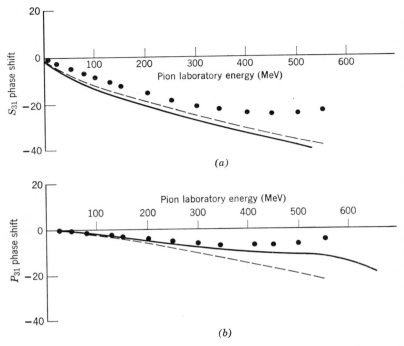

FIG. 2. $I = \frac{3}{2}, J = \frac{1}{2}$ phase shifts as a function of pion laboratory kinetic energy E_L (with apologies to Steiner[6] for the choice of variable). Here and in Figs. 2–5 (all taken from Ref. 12) the solid curve is the solution of the Frye-Warnock one-channel N/D equation (7) with the input inelastic factor η taken from Refs. 1 and 4; the dashed curve is the solution for $\eta = 1$; the dots represent the results of phase shift analysis of Ref. 1. The cutoff $\Lambda = 19.0$ and 18.6 (in units with $m_\pi = 1$) for the elastic and inelastic unitarity calculations, respectively.

was calculated in the low energy region[17] using f as determined from the phase shift analyses. A simple form was chosen to represent this discrepancy or inadequacy in B. Then $B' \equiv B + \Delta$ was used as a new input potential into the N/D equations. This improved matters but only to a limited degree.[25] The contributions to the u channel potential arising from the exchange of the $J = \frac{1}{2}$, πN resonances have been calculated and found to be quite small.[26] Perhaps a good treatment involving the exchange of Regge trajectories to determine B would improve matters. On the other hand, we note the really excellent agreement with experiment in low-energy NN scattering that Scotti and Wong[27] obtained from their N/D calculations which used similar approximation to those we have just described. This suggests that the difficulties with the πN program might lie elsewhere.

There are good indications that the crucial factor missing in the above calculations is that the inelastic channels have not been explicitly considered. In other words, important CDD effects (with respect to the one-channel πN calculations) are present. Unfortunately, except for the S_{11} wave in which it is known that the ηN channel dominates the inelastic production we know very little about the breakdown of the inelastic cross section for a given πN partial wave into quasi-two-body channels. It is quite possible that the important channel (or channels) that must be explicitly considered in a dynamical ND^{-1} calculation is a very high mass one, e.g., as emphasized by Dalitz[28] all the πN resonances may be quasi-bound states of the 3-quark system. As detailed by Rosenfeld,[29] the (enormous number of) bubble chamber pictures have been taken and the events must be measured and the relevant information extracted. The results will be extremely important. In the absence of this detailed knowledge of the inelastic partial-wave cross sections, a phenomenological approach seems to be the most fruitful way to proceed.

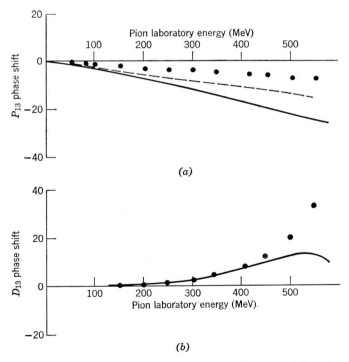

(a)

(b)

FIG. 3. $I = \frac{1}{2}, J = \frac{3}{2}$ phase shifts as a function of E_L. The cutoff $\Lambda = 18.0$ and 19.0 for the elastic and inelastic unitarity calculations. In (b), the D_{13} phase shift calculation using elastic unitarity is too small to be shown on this scale.

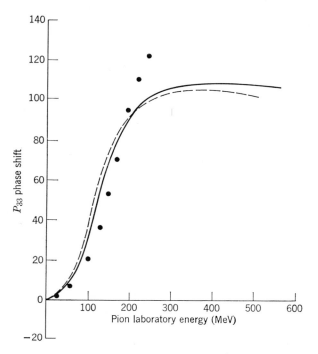

FIG. 4. The P_{33} phase shift as a function of E_L. $\Lambda = 17.\dot{3}$ and 19.0 for the elastic and inelastic calculations.

We[30] have applied a simple 1-pole 2-channel ND^{-1} model to analyze the πN partial wave phase shift results. Channel 1 is the πN channel and channel 2 is a purely phenomenological two-body channel which is supposed to represent the inelastic channels. As the input potential in the multi-channel ND^{-1} equations (6), we take a single pole at the energy W_0:

$$B_{ij} = \gamma_{ij}/(W - W_0), \qquad \gamma_{ij} = \gamma_{ji} \tag{13}$$

Making the subtraction in D at $W = W_0$, we have[31]

$$N_{ij} = \gamma_{ij}/(W - W_0) \tag{14}$$

$$D_{ij} = \delta_{ij} - \gamma_{ij}\, di(W) \tag{15}$$

with

$$d_i = \frac{W - W_0}{\pi} \int_{W_iT}^{\infty} \frac{dW' \rho_i(W')}{(W' - W_0)^2 (W' - W - i\varepsilon)} \tag{16}$$

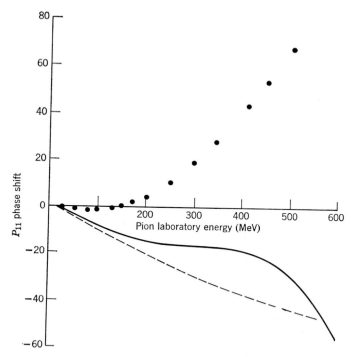

FIG. 5. P_{11} phase shift as a function of E_L where the nucleon pole is forced to appear as a dynamical bound state at the correct energy. For the elastic unitarity calculation, $\Lambda = 16.5$ giving (from the residue of the pole) an output $g^2/4\pi = 29.0$ and the computed scattering length is -0.28. For the inelastic calculation, $\Lambda = 17.2$ yielding noutput $g^2/4\pi = 25.1$ and a scattering length of -0.22.

This simple model applied to the P_{11} wave can fit the position and residue of the nucleon bound state and the detailed behavior of δ as shown in Fig. 6.[32] The fit to η is not very good. Probably a good fit to η could be obtained using a third channel without changing the nature of the solution. This model yields the following interpretation of the dynamics: In the absence of interchannel coupling (set $\gamma_{12} = 0$), each channel (πN and the phenomenological second channel which was taken to be "σ" N) has a bound state or low-lying resonance. They are fairly close in energy and as the coupling is turned on, the lower of the two poles moves to lower energy (the position of the nucleon pole) and the higher energy pole moves up (to the position of the Roper resonance).

The detailed shape of the P_{33} resonance is easily fit by this model. However, the dominant force is the diagonal interaction in the inelastic channel and the $N_{33}{}^*$ appears in a quasi-bound state or CDD effect with

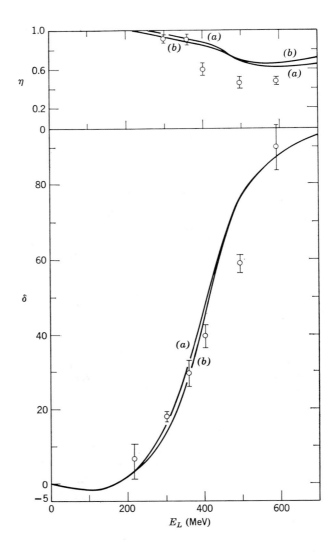

FIG. 6. Plots of calculated values (solid curves) of η and δ for the P_{11} partial wave versus E_L using the two-channel, 1-input pole model [Eqs. (13)–(16)]. This figure is taken from Ref. 30. The calculated solid curves (a) and (b) corresponding to two different choices for the phase space factors fall on top of the low-energy, energy-dependent phase shift analysis of Ref. 1 for $E_L < 200$ MeV. The experimental points shown are from Ref. 4. Very good output values for $g^2/4\pi$ are obtained.[32]

respect to the πN channel. This fit to the data is shown in Fig. 7 along with the results of a one-channel calculation which approximates the input B by 2 poles. This one-channel model does not give a good fit above the resonance (and we note that the nearly input pole in B is repulsive). Furthermore, the two channel model yields a satisfactory fit to the in-elastic factor η.

Although these results for the P_{11} and P_{33} partial waves are quite suggestive, we should bear in mind that clearly in each partial wave with enough input poles in the πN channel, we would find solutions (in addition to those in Figs. 6 and 7) which would fit all the present data in a manner

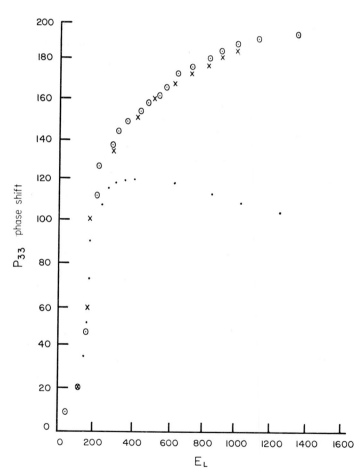

FIG. 7a. See page 128 for caption.

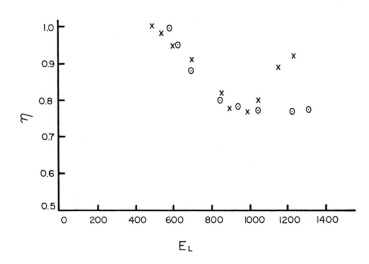

FIGS. 7a (page 127) and 7b (above). Plots of calculated values of δ and η for the P_{33} partial wave versus E_L. The $\bigcirc$'s correspond to the results of a two-channel, 1-input pole calculation using (13)–(16) with $W_0 = 5.0$, $\gamma_{11} = 89.5$, $\gamma_{12} = 233$, $\gamma_{22} = 815$, and $W_2{}^T = 10.7$. The ρ_i were taken as $k_i{}^3/W^3$. The •'s correspond to the results of a one-channel, 2-input pole calculation with input strengths of -1150 and 4700 and positions -13 and -618. The $\times$'s correspond to the results of the phase shift analysis of Ref. 4. $\otimes$ stands for $\bigcirc + \times + •$.

such that there are no CDD effects with respect to the πN channel. Furthermore, even if the above physical interpretations of P_{11} and P_{33} partial waves obtained from the two-channel calculations are essentially correct, we cannot conclude that the Chew-Low reciprocal bootstrap results[23] were totally fortuitous. The bootstrap equations only gave relations between the residues. Thus, e.g., even though the forces determining the position of the $N_{33}*$ *might* come from a second channel, the forces in the πN channel might have considerable effect in determining the residue. However, we stress that calculations of the type which simply (by use of crossing) find in which partial wave a given πN resonance produces a "large" u channel exchange force are essentially useless in understanding the dynamics of πN scattering.

Reasonably good fits to most of the resonant amplitudes are obtained using the two-channel model. The most notable exceptions are the S-waves. Perhaps a second input pole or a third channel is needed here.

We would like to suggest that this multichannel parametrization would be useful to use in a direct (perhaps after one has made an energy independent analysis) analysis of the data. Each partial wave amplitude in the

1-pole, 2-channel approximation is a simple function of only five parameters making a best fit search analysis easy. The parameters can be physically interpreted: three are related to the strength of the input forces, the input pole position is related to the range of interaction and the effective threshold of the phenomenological inelastic channel is the fifth parameter. More input poles or inelastic channels can be easily added. An analysis of the elastic data together with the large number of inelastic bubble chamber events[29] using such an extended model in which the inelastic channels are taken to be $\pi N_{33}{}^*$, σN, ρN, etc. should prove very useful. We are now examining such a parametrization.

Finally we remark that in analogy with the analysis of NN scattering in which the high J waves are represented by π exchange,[33] it might be useful to supplement any πN analyses of the low J waves with the t channel $\pi\pi \to $ " σ " $\to N\overline{N}$ term and u channel N exchange to represent the high J waves.

I would like to thank P. Coulter for many helpful discussions.

References

1. L. Roper, *Phys. Rev. Letters*, **12**, 340 (1964); L. Roper and R. Wright, *Phys. Rev.*, **138**, B921 (1965).
2. B. Bransden, P. O'Donnell, and R. Moorhouse, *Phys. Letters*, **11**, 339 (1964); *Phys. Rev.* **139**, B.1566 (1965).
3. P. Auvil, C. Lovelace, A. Donnachie, and A. Lea, *Phys. Letters*, **12**, 76 (1964); A. Donnachie, A. Lea, and C. Lovelace, ibid., **19**, 146 (1965); C. Lovelace, CERN Report No. Th-837 October, 1967.
4. P. Bareyre, C. Bricman, A. Stirling, and G. Villet, *Phys. Letters*, **18**, 342 (1965); P. Bareyre, C. Bricman, and G. Villet, *Phys. Rev.*, **165**, 1730 (1968).
5. C. Johnson, Lawrence Radiation Laboratory Report UCRL-17683, August, 1967.
6. C. Johnson and H. Steiner, this volume.
7. C. Lovelace, this volume.
8. Lovelace (Ref. 7) defines positions and widths of the higher mass inelastic resonances with respect to a minimum in η. Some further study is probably necessary in examining his definition of a resonance.
9. M. Froissart, *Nuovo Cimento*, **12**, 191 (1961).
10. G. Frye and R. Warnock, *Phys. Rev.*, **130**, 478 (1963).
11. J. Fulco, G. Shaw, and D. Wong, *Phys. Rev.*, **138**, B.702 (1965).
12. P. Coulter and G. Shaw, *Phys. Rev.*, **141**, 1419 (1966).
13. J. Hamilton, P. Menotti, G. Oades, and L. Vick, *Phys. Rev.*, **128**, 1881 (1962).
14. J. Hamilton and T. Spearman, *Ann. Phys. (N.Y.)*, **12**, 172 (1961); J. Hamilton and W. Woolcock, *Rev. Mod. Phys.*, **35**, 737 (1963); A. Donnachie and J. Hamilton, *Phys. Rev.*, **133**, B1053 (1964); **138**, B678 (1965). The generalized potentials B considered by Hamilton and co-workers cutoff the far-away contributions in Eq. (3).
15. L. Castillejo, R. Dalitz, and F. Dyson, *Phys. Rev.*, **101**, 453 (1956).
16. See, e.g., P. Coulter, R. Garg, and G. Shaw, *Phys. Rev.*, **166**, 1708 (1968) and references contained there.

17. P. Coulter and G. Shaw, *Phys. Rev. Letters*, **19**, 862 (1967).
18. E. Abers and C. Zemach, *Phys. Rev.*, **131**, 2305 (1963).
19. J. Ball and D. Wong, *Phys. Rev.*, **133**, B179 (1964); **138**, AB4(E) (1965).
20. J. Ball and D. Wong, *Phys. Rev.*, **130**, 2112 (1963).
21. T. Spearman, *Phys. Rev.*, **129**, 1847 (1963).
22. As noted by Sakurai (*Phys. Rev. Letters*, **17**, 552 (1966)) the ρ-dominance model gives the correct S-wave πN scattering lengths. However, this model does not have the correct energy dependence.
23. G. Chew, *Phys. Rev. Letters.*, **9**, 233 (1963), F. Low, ibid., **9**, 279 (1962).
24. See Fig. 6 of Ref. 12. Alternatively, a good fit could be obtained by inserting the Roper resonance as a CDD effect with respect to the one-channel N/D equations.
25. As mentioned above, a cutoff in the N/D equations can compensate in B for a slowly varying function of W.
26. P. Coulter and G. Shaw, *Phys. Rev.*, **153**, 1591 (1967). Note that considering the detailed shape of the P_{33} resonance reduces its contribution to the exchange potential by a factor of 0.75 as compared with the narrow-width approximation.
27. A. Scotti and D. Wong, *Phys. Rev.*, **138**, B145 (1965).
28. R. Dalitz, this volume.
29. A. Rosenfeld and P. Söding, this volume.
30. J. Ball, G. Shaw, and D. Wong, *Phys. Rev.*, **155**, 1725 (1967); J. Ball, R. Garg, and G. Shaw, *Phys. Rev.*, (in press).
31. The $-W$ region is ignored.
32. The 3γ's are fixed by requiring that the nucleon pole, the zero in δ and the energy at which $\delta = 90°$ occur at the correct positions.
33. M. MacGregor, M. Morarvcsik, and H. Noyes, *Phys. Rev.*, **123**, 1835 (1961).

Structure of the Differential Cross Section at High Energy

MARC ROSS

University of Michigan, Ann Arbor, Michigan

There is rich structure in the high energy πN differential cross section which can be qualitatively understood in terms of two kinds of simple physical processes. I would like to discuss these processes, mentioning, briefly, the sometimes successful and sometimes quite forced Regge formalism which is also used to describe the structure. The processes (Regge formalism) I will discuss very briefly are (*A*) multiple scattering among the parts of composite particles (related to cuts) and (*B*) *S*-channel resonances, especially the conjectured "new bootstrap" between them and exchanged trajectories (related to the pole terms and background integral).

First let us look at the data. In Fig. 1 the forward scattering is shown. Note that the secondary maximum in charge exchange scattering is comparable with the proper secondary maximum in elastic scattering. This maximum goes roughly as $S^{-2.5}$. At high energy we see that the secondary maximum is still visible in charge exchange (whose 0° cross section goes roughly as S^{-1}) but has vanished from the elastic. Instead, in this general momentum transfer region, we have a break in the differential cross section which appears, on the basis of scant data, to be essentially S independent like the forward cross section. This break seems to be a phenomenon of different origin than the secondary maximum.

In the backward direction (Fig. 2) we have a sharp $\pi^+ p$ backward peak with secondary maximum and a smaller rather flat $\pi^- p$ backward peak (see especially the recent Cornell experiment reported by H. White). These cross sections scale roughly as $S^{-2.5}$ (with some change in shape).

A. Multiple Scattering

Qualitatively, due to the composite structure of hadrons, there should be multiple scattering of the components (in the sense of the impulse approximation or the Glauber correction for reactions on deuterium). If the

131

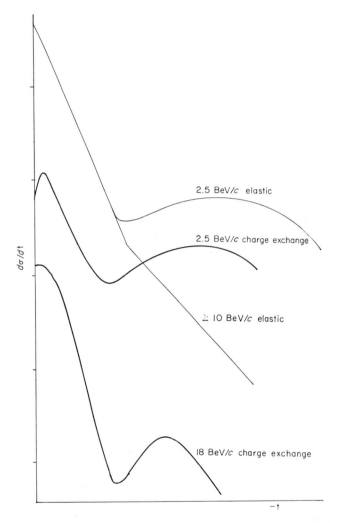

FIG. 1. Forward angular distribution in pion–nucleon elastic and charge exchange scattering.

single scattering amplitude of a particle on a component part of another particle (Fig. 3a) is, for example,

$$Ae^{Ct/2} \rightarrow Ae^{Bt/2}$$

average over the system

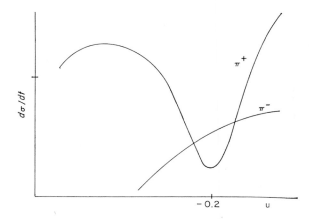

FIG. 2. Typical backward elastic angular distributions at high energy.

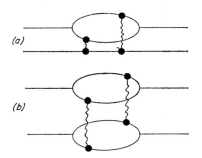

FIG. 3. (a) Scattering of a particle on a composite particle (double scattering illustrated). The interaction between target and projectile is indicated by the wavy line. (b) Double scattering diagram of the type shown by Mandelstam to give a " persistent " contribution at arbitrarily high energy [as opposed to (a)]. The proof involves detailed dependence on behavior of off-mass Feynman propagators.

then there is "double scattering" involving primarily diffraction scattering on a second component, which goes like

$$\int Ae^{Ct'/2}f(t')i\pi\,\delta(E)[i(k/4\pi)\sigma_T\,e^{C't''/2}]\propto -AP(t)e^{\bar{C}t/2}$$

This amplitude is negative and broader in t with respect to the single scattering amplitude. It is negative because it is a shadowing correction. That it is broader can be understood in terms of the spatial alignment needed to obtain large double scattering. High energy (1.7 BeV/c) $p - \alpha$

scattering shows the effects of double and triple scattering in the differential cross section (Fig. 4). This is part of a beautiful experiment by Palevsky and collaborators[1] at the Cosmotron. The multiple scattering interpretation was made by Bassel and Wilkin.[2] Note that simple diffraction of a wave by a homogeneous absorber is *not* a quantitative way to describe this data. However $p - C$ and $p - O$ data from the same experiment shows this description (diffraction of a wave by a homogeneous optical potential) becomes fairly accurate where there are enough components (i.e., $A \gg 1$), since for these nuclei it agrees reasonably well with the multiple scattering calculation.

Quantitatively we are in trouble when we attempt to calculate these processes in particle physics. We find several difficulties: (*1*) For many purposes a detailed model of the composite structure is needed. (*2*) There is a problem in counting the contributions. It is stated[3] on the basis of a paper by Mandelstam[4] that only processes of type (*b*) (Fig. 3), as opposed to type (*a*), persists at high energy. There are reasons to question this conclusion. (*3*) Inelastic reactions are probably quite important in the multiple scattering. Probably the direction to pursue is to find effects which

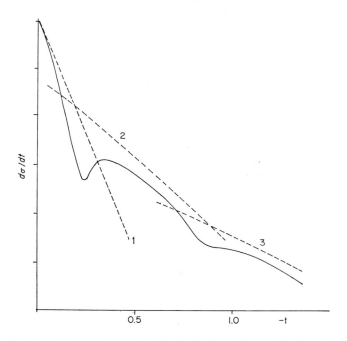

FIG. 4. Sketch of p–α elastic scattering. Dashed curves show qualitatively the separate orders of multiple scattering.

are relatively model independent such as, perhaps, polarization, crossed channel partial-wave analysis, 2–3 particle processes.

The evidence for multiple scattering (or cuts) in particle physics is largely in pion reactions. Perhaps the best evidence, since it is a structure characteristic of double scattering, are the fixed breaks in the differential cross section. This effect has been considered by several people.[5] Since the double scattering is associated with diffraction, the S dependence should be essentially the same as that of single scattering. The decreasing secondary maximum is presumably not connected with this process. At higher energy we expect to see the fixed break come into charge-exchange scattering. Other effects which have been attributed to multiple scattering (or cuts) are the crossover of $\pi^- p$ and $\pi^+ p$ near $0°$,[6] polarization in charge-exchange scattering,[7] corrections to forward (quark model) cross-section relations.[8]

The crossover is also an effect characteristic of multiple scattering for the following reason: The grayness or opacity of $\pi^+ p$ scattering as a function of impact parameter is shown by a typical curve (Fig. 5). For $\pi^- p$ the total cross section is larger and presumably the shape is roughly similar. We obtain the dashed curve of Fig. 5 (exaggerated) which has a larger effective radius. Thus the differential cross section goes roughly as

$$\frac{d\sigma}{dt} \propto (\sigma_T)^2 e^{R^2 t} \qquad \text{and} \qquad \sigma_T \propto R^2$$

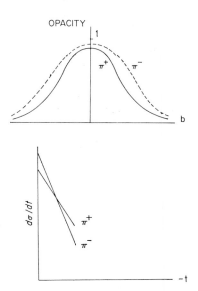

FIG. 5. (a) Opacity versus impact parameter for $\pi^- p$ and $\pi^+ p$. (b) Crossover of forward $\pi^- p$ and $\pi^+ p$ differential cross sections.

(In practice R^2 does not increase as rapidly as σ^T) so that the differential cross sections look as in Fig. 4b. The increased shadowing effect thus naturally explains the crossover. Possible Regge language (conspiring cuts) for this process is briefly discussed by Barger and Durand.[6]

Let me conclude these remarks by stating my belief that we should deduce from such clear semiclassical explanations the Regge formalism appropriate to the physical process.

B. S-Channel Resonances and the New Bootstrap

Schmid and collaborators[9] have recently interpreted so-called finite energy sum rules to mean that an energy averaged imaginary, and perhaps real, amplitude at low energies has the same properties, especially as a function of momentum transfer, as, let us say, the trajectory exchange amplitude which dominates at high energy. The sum rule, for any n, reads

$$\int_0^N dv\, v^n \, \text{Im}\, F(v) = \sum \frac{\beta_i N^{\alpha_i + 1 + n}}{(\alpha_i + 1 + n)\Gamma(1 + \alpha_i)}$$

where

$$\sum_i \frac{\beta_i v^{\alpha_i}}{\Gamma(1 + \alpha_i)} + B(v)$$

is the Regge analyzed imaginary amplitude (without cuts) where B is the background term. They suggest that one choose the energy N roughly at the boundary between "resonance" and "exchange" regions. The "new bootstrap" interpretation can be described by saying that these general analyticity relations can be satisfied by rewriting them ($v < N$) as

$$\overline{\text{Im}\, F(v)} = {\sum}' \frac{\beta_i v^{\alpha_i}}{\Gamma(1 + \alpha_i)}$$

where the left-hand side is a rather narrow energy average and the right-hand side is the sum over leading trajectories; which implies that the low energy resonance amplitude oscillates about the continuation of the high energy trajectory exchange amplitude. I find this interpretation quite exciting but meaningful at low energy only if the analytic form of the Regge amplitude is known and, in particular, if it is dominated by a few leading trajectories. The trouble is other cuts and subsidiary trajectories are poorly determined at low energy. In this interpretation only fluctuation terms of average value zero are contained in B. Let me emphasize again that the reservations I have relate to the way Schmid and collaborators wish to satisfy the sum rules. The sum rules are very general; they do not require this particular solution. The experimental evidence advanced for this interpretation is, in spite of these problems, rather good. It is essentially that the low energy charge exchange amplitude has the same qualitative properties when appropriately averaged over energy as the high energy amplitude which is dominated by a spin-flip term which vanishes near $t = -0.6$.

If we assume resonance dominance of the amplitude, this interpretation constitutes an approximate equivalence of the S-channel resonance amplitude and the trajectory exchange amplitude:

$$\sum \frac{\gamma_i \gamma_j}{S_0 - S - iS^{1/2}\Gamma} P_l(\cos \theta) \approx \sum \frac{\beta v^\alpha}{\Gamma(1 + \alpha)} \frac{(\pm 1 - e^{-i\pi\alpha})}{\sin \pi\alpha}$$

(Note: fluctuations are missing from the right-hand side where we consider only leading trajectories. The trajectories and their analytic forms for $\alpha < -1$ are quite uncertain.) This postulated relation between low energy resonances and leading exchange trajectories suggests a new kind of bootstrap: The same trajectories may occur in both direct and exchange channel.

Dikmen and I have independently been considering an extended version of the approximate equivalence of direct channel resonance and exchange amplitudes. We assume resonance dominance of the non-diffractive *high energy* scattering. One finds an immediate suggestion of a new bootstrap in backward scattering. On the one hand, backward $\pi^+ p$ and $\pi^- p$ are readily described in terms of N and N_{33}^* trajectory exchange; on the other hand, the large $\pi^+ p$ backward peak can be associated with a strong excess (as weighted from the πN channel) of one parity over the other. The "bootstrap" is thus: low weight of $I = \frac{3}{2}$ negative parity N^*'s $\leftrightarrows$ large N trajectory exchange contributions of $\pi^+ p$ backward. The $\pi^- p$ backward is associated with even distributions of different parity N^*'s. I do not know how to carry out this bootstrap quantitatively. The idea is qualitatively confirmed, however, by recent compilations (see talks of Lovelace and Steiner) of lower energy resonances, in that three of the four I-spin parity states are strong in the πN channel and $I = \frac{3}{2}$, $P = -1$ is negligible. In particular up to $M = 2$ BeV the weights are

	$I = \frac{1}{2}$		$I = \frac{3}{2}$	
	$P = 1$	$P = -1$	$P = 1$	$P = -1$
$\sum \left(j + \frac{1}{2}\right)\dfrac{\Gamma_{el}}{\Gamma_t}$	4.1	3.2	5.0	0.6

(I would like to thank G. Kane for this suggestion.)

Moreover the π^+p and π^-p backward scattering have been described at moderate energies (by Barger and Cline[10] and by Dikmen,[11] respectively) entirely in terms of " known " S-channel resonances. We have extended this idea to higher energies with an ultrasimple model of resonances with parameters drawn at random from distributions skewed in l and j. A result averaged over many samples is shown for π^-p at 6 BeV/c with a density of 100 resonances/BeV of mass,[12] with l chosen from a uniform distribution cutoff at 7 and $j = l + \frac{1}{2}$ for $l = l_{max}$, otherwise uniform (Fig. 6).

In the spirit of the approximate equivalence of high energy resonance and exchange amplitudes we find that the maximum angular momentum N^*'s of significant weight in the πN channel determine the position of the minima in near-forward and near-backward (for π^+p) directions. In a very simple model we obtain $l_{max} = \frac{1}{4}(M - M_N)$, rather than $l_{max} \propto M^2$. The reason for this is that the dip in the differential cross section occurs at constant momentum transfer, i.e., at $-t^{1/2} \approx k\theta \approx$ const. (Similarly for the backward dip in π^+p.) Now the first dip of a P_l is at $\theta \propto 1/l$; thus the states of weight in the πN channel must satisfy $l \propto k \propto M$ (which is classically satisfying). The bumps actually seen in πN total cross section are probably the states of greatest weight. The simplest model will yield $l_{max} \propto M$. Since this result depends on detail on our simple model we do not take it seriously, but advance it to show some potentialities of the " new bootstrap " hypothesis.

I conclude by emphasizing two things:

1. The essential question is the possible dominance by resonances of a significant part of the amplitude at high energy (such as the whole amplitude or as I have discussed here, the nondiffractive part). It is of little interest if the resonance amplitude is so smooth in energy that the resonance description constitutes simply a smooth partial wave analysis. The test of the idea is in observing energy fluctuations in high energy cross sections. At present the very limited data on fluctuations is consistent with an interesting resonance model.

2. If there is a dominant resonance amplitude the resonance parameters must be *highly organized*. This is most evident when one considers inelastic processes: e.g., compare $\pi^-p \to K^+\Sigma^-$, which is very small, and $\pi^-p \to K^\circ\Sigma^\circ$. Single resonances would contribute about equally to these processes. Thus, in general, arguments made on the basis of random parameter values will not be valid. We hope that energy fluctuations in the elastic differential cross section might be insensitive to the difference between the highly organized resonance and random resonance models so that we could make sensible predictions for this case. The fact that resonances, if important, must be organized makes the " new bootstrap " potentially very rich.

I would like to thank many participants for remarks which have been incorporated in this written version of my talk.

References

This is a brief survey of references. The author apologizes for the many omissions.

1. G. Bennett et al., *Phys. Rev. Letters*, **19**, 387 (1967).
2. R. Bassel and C. Wilkin, *Phys. Rev. Letters*, **18**, 871 (1967); see also E. Coleman and V. Franco, *Phys. Rev. Letters*, **17**, 827 (1966).
3. Abers, Burkhardt, Teplitz, and Wilkin, *Nuovo Cimento*, **27**, 365 (1966).
4. S. Mandelstam, *Nuovo Cimento*, **30**, 1131, 1117, 1148 (9163).
5. Harrington and Pagnamenta, *Phys. Rev. Letters*, **18**, 1147 (1967); Huang, Jones, and Teplitz, *Phys. Rev. Letters*, **18**, 146 (1967); Deloff, *Nucl. Phys.*, **B2**, 597 (1967); J. Pumplin, private communication; E. Schrauner, private communication.
6. V. Barger and L. Durand, *Phys. Rev. Letters*, **19**, 1295 (1967).
7. V. N. de Lany et al., *Phys. Rev. Letters*, **18**, 148 (1967).
8. R. Heinz and P. De Souza, private communication.
9. Dolen, Horn, and Schmid, *Phys, Rev. Letters*, **19**, 402 (1967) and Cal. Tech. Report.
10. V. Barger and D. Cline, *Phys. Letters*, in press.
11. F. N. Dikmen, *Phys. Rev. Letters*, **18**, 798 (1967).
12. The number 100 is based on the number of resonances observed at low energy augmented by the additional ones which should be present at high energy by non-sense channel arguments and by the introduction of new trajectories as the mass increases.

Regge Poles and High Energy πN Scattering*

STEVEN FRAUTSCHI†

California Institute of Technology
Pasadena, California

We have heard about the understanding of πN scattering which has been gained by very detailed work on fitting phase shifts in the direct channel partial wave expansion at low energy (low s). In this paper I shall review the analogous attempts to understand higher energy scattering by means of Regge poles. In this work, to fit forward angle scattering, one uses the Regge pole expansion for the crossed (t) channel, which can be obtained by analytic continuation of the t-channel partial wave expansion. To fit backward angle scattering, one uses the Regge pole expansion for the crossed (u) channel, which can be obtained by analytic continuation of the u-channel partial wave expansion.

To put things in proper perspective, I should immediately emphasize that the Regge fits to high energy data are still in the stage of using two or three exchanged objects to explain a small number of peaks, dips, and bumps. This is analogous to phase shift fitting in the days when we only had three peaks, to which we attached three resonances, with some uncertainty about the spin-parity assignments of all but the first resonance. Some of the gaps in the Regge fits are experimental; the high energy polarization data are still quite incomplete. On the theoretical side, whereas one has only to determine how many partial waves of fixed $l = 0, 1, 2, \ldots$ to keep in the phase shift analysis, the Regge pole fitter is by no means set when he decides how many poles to include; he must also determine what their variable $\alpha(t)$ is, whether to include the accompanying "Regge cuts" (usually ignored), and whether to include the fixed poles recently proposed by Jones and Teplitz,[1] and Mandelstam and Wang[2] (usually ignored). For all these reasons, the uniqueness of the existing fits is highly questionable. Therefore, I shall not attempt to evaluate the accuracy of what has been done, but will content myself with a review of the structural features

* Work supported in part by the U.S. Atomic Energy Commission. Prepared under Contract AT(11-1)-68 for the San Francisco Operations Office, U.S. Atomic Energy Commission.

† Alfred P. Sloan Foundation Fellow.

which stand out, much as the "first, second, and third" resonances stood out in the low-energy scattering.

In the phase shift analysis, there were two different approaches; both started with the partial wave analysis at each energy, and both used some kind of "smoothing" between energies, but the CERN group made explicit use of analyticity to help in the "smoothing." A somewhat similar division exists in approaches to Regge poles.

First, there is the direct comparison of Regge pole models with the high-energy data. Analyticity is implicit here, but is only used in the high-energy region.

The second method leans more heavily on analyticity, using fixed momentum transfer dispersion relations to connect the high-energy (Regge) parameters to the data at lower energy. Recently this second approach has been developed[3-5] into the Finite Energy Sum Rules (dubbed FESR), which express systematically the idea that the high-energy behavior is already prefigured in the lower energy behavior. Specifically, the FESR says that for an even or odd πN scattering amplitude F which is expandable as a sum of Regge poles at high energies,

$$S_n = \frac{1}{N^{n+1}} \int_0^N v^n \, \mathrm{Im} \, F(v) \, dv = \sum_i \frac{\beta_i N^{\alpha_i}}{(\alpha_i + n + 1)\Gamma(\alpha_i + 1)} \tag{1}$$

where $v = (s - u)/2M$ is an energy variable, and the momentum transfer t is held fixed. The integration in Eq. (1) is defined over the right-hand cut in s and includes the Born term even if it occurs at negative v. The left-hand side involves an integral over the imaginary part of the amplitude, weighted by v^n, up to the energy corresponding to $v = N$; the right-hand side is a sum over Regge terms evaluated at $v = N$. Thus, inserting the low-energy data into the left-hand side, we can use it to give information about the Regge parameters β and α.

In practice, mixtures of these two methods are often used. For example, Rarita et al.[6] use certain sum rule constraints in the course of their work on direct fits at high energies, and Dolen et al.[5] use information from high-energy fits in some of their applications of FESR.

The simplest case to analyze is $\pi^- p \to \pi^0 n$ at small t, since only $I = 1$, $S = 0$ exchanges with $G = +$ can contribute to this reaction, and the ρ meson is the only known particle with these attributes. Because of the simple form the Regge model takes in this case, $\pi^- p \to \pi^0 n$ provides one of the best tests that the model is on the right track. The main aspects of the test, as worked out by Logan,[7] Rarita and Phillips,[8] Höhler et al.,[9] and Arbab and Chiu.[10] are as follows:

Shrinking Peak. If we make the one-pole approximation (pure ρ

trajectory exchange), then the cross section is expected to behave like

$$\frac{d\sigma}{dt} = f(t)\left(\frac{s}{s_0}\right)^{2(\alpha_\rho(t)-1)} \tag{2}$$

with the power $\alpha_\rho(t)$ decreasing as t becomes negative, leading to a shrinking peak. While this shrinkage has not been manifest in diffraction peaks, where several exchanges are present and the dominant (Pomeranchuk) exchanges may have an unusually flat $\alpha(t)$, the shrinkage does show up clearly in $\pi^- p$ charge exchange. You can see it in Fig. 1, in the *spacing* between $d\sigma/dt$ measured[11,12] at different energies, which becomes greater at large t.

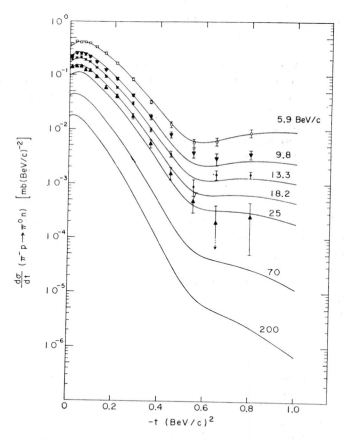

FIG. 1. $\pi^- p \to \pi^0 n$ differential cross sections at 5.9, 9.8, 13.3, and 18.2 BeV/c compared to a ρ exchange model of Rarita et al.[6]

Connection to ρ Particle. By comparing formula (2) to the data as summarized in Fig. 1, one deduces $\alpha_\rho(t)$. It is about 0.55 at $t = 0$, and falls in a roughly linear fashion to zero at about $t = -0.6$ BeV^2. Continuation of the line defined in this fashion to positive t brings α_ρ to 1 at about $t = 0.5$ BeV^2, where the ρ particle is actually observed. These conclusions, first reached by high-energy fits,[7-10] also emerge from the FESR.[5]

Explanation of Dip at $t = -0.6$ BeV^2. It has been believed for some time that the ρ exchange amplitude is predominantly helicity flip rather than non-flip, except where special factors intervene. In this picture, the forward dip in πN charge exchange (Fig. 1) is due to the kinematic zero which prevents helicity flip in the forward direction (a consequence of angular momentum conservation). The dip observed near $t = -0.6$ BeV^2 is attributed[8-10] to a zero occurring in the helicity flip amplitude when the exchanged spin (α_ρ) passes through zero.

The consistency of this picture with our other high-energy information is seen from the fact that the value of $t = -0.6$ BeV^2 at which $\alpha_\rho = 0$ in this picture agrees well with the determination of α_ρ by means of the energy dependence of $d\sigma/dt$. However, Mandelstam and Wang[2] and Jones and Teplitz[1] have pointed out that fixed poles occur at wrong signature nonsense points. Now $\alpha = 0$ in the helicity flip amplitude is such a point ("nonsense" points are unphysical points, i.e., $\alpha = 0, -1, -2, \ldots$ for the helicity flip amplitude and $\alpha = -1, -2, -3, \ldots$ for the non-flip amplitude. Since the ρ trajectory can make particles at $\alpha = 1, 3, 5, \ldots$, its "wrong signature" points are $\alpha = 0, 2, 4$, and also by extension the negative even integers.) The fixed pole can cancel the helicity flip zero at $\alpha_\rho(t) = 0$. Thus the dip at $t = -0.6$ BeV^2 is not a fundamental test of the Regge pole hypothesis, but an indication that the effect of fixed poles may be rather small. As Mandelstam and Wang discuss, this places fixed poles in a similar category to Regge cuts—both are necessary on theoretical grounds when a third double spectral function is present, but both may have rather small coefficients in practice.

Aside from the possibility of fixed poles, another question mark in the interpretation just advanced has been the lack of a direct verification that it is the helicity flip amplitude which has the zero at $t = -0.6$ BeV^2. Here the FESR have helped recently[5] by confirming that the helicity non-flip amplitude is small throughout the region under consideration, whereas the helicity flip amplitude is big except at $t \approx -0.6$ BeV^2. The way in which this behavior emerges from the integral over low-energy resonances in the FESR is that the resonance contributions to the non-flip amplitude fluctuate about a *small* average value, whereas the resonance contributions to the flip amplitude tend to add up with the same sign, producing a *large* average value. The contribution from resonances also varies with angle, of course, and in nearly all the main resonance contributions to the flip amplitude, the angular dependence consists of the flip zero at $t = 0$, then a maximum, followed by a node near $t = -0.6$ BeV^2 thus accounting for the helicity flip zero at $t \approx -0.6$ BeV^2.

Charge-Exchange Polarization. In the pure ρ exchange model, both flip and non-flip amplitudes have the same phase, which implies zero polarization. Actually there is some experimental indication that the charge-exchange polarization P does not vanish—Bonamy et al.[13] obtain $P \sim 15\%$ over the range 0.24 BeV$^2 \geq |t| \geq 0.04$ BeV^2 at both 5.9 BeV/c and 11.2 BeV/c. Additional terms as small as $\sim 10\%$ of ρ exchange could account for this amount of polarization provided they were suitably out of phase, but if the polarization really turns out to vary so slowly with energy, interesting problems will be raised.

Several types of explanation for a nearly energy independent P have been proposed:

(i) A second trajectory (ρ'),[14–17] lying not far below ρ.

(ii) A Regge cut.[18–19] In particular, the cut due to $(\rho + P)$ exchange lies as high as ρ itself and therefore could easily explain the energy dependence.

(iii) Contributions from direct channel resonances.[20–23] Of course, any particular contribution of this type can alternatively be expressed in the language of the t-channel J plane; the tail of a low-energy Breit-Wigner resonance,[20,21] for example, would fall off like s^{-1}, thus behaving like a fixed pole with $\alpha = -1$.

In detailed attempts to fit the data, various investigators[5,15,17] obtain fits with a second pole, ~ 0.4 lower than the ρ, in the helicity flip amplitude, It is not possible at present to tell whether this is a cut or a pole—especially since the contribution from a cut is so dependent on the unknown details of the discontinuity across the cut.

Elastic $\pi^+ p$ and $\pi^- p$ Scattering. The theory of these reactions at small t is more complicated, because both $I = 0$ and $I = 1$ states can be exchanged. At least three trajectories, P, P', and ρ, are required. The types of data available are: (i) total cross sections; (ii) the phase of the forward amplitudes, from coulomb interference measurements; (iii) differential cross sections; and (iv) polarizations. These data, in conjunction with related data on $pp \to pp$, have been intensively studied in the three Regge pole model by Phillips et al.[6,8] at Berkeley.

Total Cross Section and Phase of Forward Amplitude. $I = 0$ exchange (P and P' exchange in the 3-pole model) contributes the same amounts to $\pi^+ p \to \pi^+ p$ and $\pi^- p \to \pi^- p$; $I = 1$ exchange (ρ in the 3-pole model) contributes opposite amounts. Since $\pi^- p \to \pi^0 n$ is pure $I = 1$ exchange, it can be used to predict the *difference* between the $\pi^+ p$ and $\pi^- p$ total cross section, and the difference between the phases of the forward amplitudes for $\pi^+ p$ and $\pi^- p$ elastic scattering. This is tested only rather crudely by the data, as described by Lindenbaum.

Turning to the *sum* of the $\pi^+ p$ and $\pi^- p$, we note that P' exchange was first introduced by Igi[24] to account for data on the forward amplitude. At the time it was somewhat disturbing to be forced to introduce a new parameter associated with no known particle exchange, but subsequently the f mesons ($I = 0$, $G = +$, $S = 0$, $J^p = 2^+$) have been discovered and fit in naturally with P' exchange In conventional Regge fits, one takes $\alpha_\rho(0) \simeq 1$ and $\alpha_{\rho'}(0) \sim 0.55$ to 0.7 to give the observed decrease in total cross section.[25] As emphasized by Lindenbaum, the correct ratio of imaginary to real parts of the forward amplitude follows automatically once one has an adequate representation of the energy dependence of the total cross section, expressed in analytic form.

Differential Cross Sections (Fig. 2). The first outstanding feature of the $\pi^+ p$ and $\pi^- p$ elastic differential cross sections is the non-shrinking diffraction peak.[26] This, together with the behavior of the Kp and $p\bar{p}$ diffraction peaks, has forced people[8,6] to assign a very small slope to P, which contrasts with the slope $d\alpha/dt \sim 1/\mathrm{BeV}^2$ found for ρ, P', and other meson and baryon trajectories.

The next feature of note is the "cross over" of $d\sigma/dt(\pi^- p)$ and $d\sigma/dt$ $(\pi^+ p)$ at $t \approx -0.2\ \mathrm{BeV}^2$, $d\sigma/dt(\pi^- p)$ being slightly larger at $t = 0$, smaller at

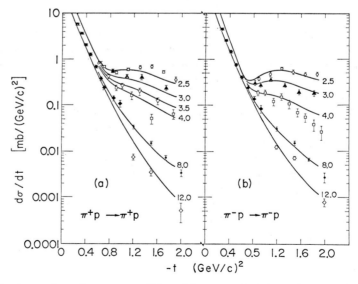

FIG. 2. $\pi^+ p \to \pi^+ p$ and $\pi^- p \to \pi^- p$ differential cross sections (taken from Chiu et al.[31]).

large t.[26] This crossover must be associated with the $C = -$, $I = 1$ exchanges which contribute with opposite sign to $\pi^- p$ and $\pi^+ p$, interfering (presumably in the non-flip amplitude) with the dominant $C = +$ exchanges. Thus the zero should also appear in $\pi^- p \rightarrow \pi^0 n$, which is pure $I = 1$ exchange, and indeed Dolen et al.[5] do find a zero in the non-flip charge exchange amplitude at nearby the same t. This zero occurs in the FESR because the contributions of dominant resonances to the low-energy non-flip amplitude have their first node near this point.

As Ross pointed out yesterday, the crossover of the diffraction peaks can be interpreted very simply—$\sigma^{tot}(\pi^- p)$ is larger, so the $\pi^- p$ differential cross section is higher at $t = 0$, but the larger σ^{tot} implies a greater radius, hence a sharper diffraction peak. There is no comparably physical interpretation in Regge language—proposals include a zero in the non-flip ρ exchange amplitude,[8,6] or conspiring poles or cuts.[27]

Another feature of small angle elastic scattering which has attracted special attention is the dip at $t = -0.6$ BeV^2, and the secondary peak, which appears at low energies but melts away at high energies (Fig. 2).[26,28,29] It is tempting to try to relate this to the dip and secondary peak in charge exchange. The ρ exchange amplitude, already known from charge exchange, turns out to have insufficient strength to explain the secondary peak in the elastic data. The next possibility is that the dip is somehow associated with the passage of $\alpha_{p'}$ through zero[30]—which can plausibly occur near $t = -0.6$ BeV^2 if $\alpha_{p'}$ follows an approximately straight line from $\alpha_{p'} = 2$ at the f meson through $\alpha_{p'} = 0.7$ at $t = 0$. The secondary peak would then *shrink* with energy because $\alpha_{p'} < \alpha_p$. This has been studied in detail by Chiu et al.[31] who find that the data can be fit in this way. There are various possible behaviors at $\alpha_{p'} = 0$:

(i) $F_{non-flip} \neq 0$, $F_{flip} \neq 0$ (the " Gell-Mann " mechanism[32])

(ii) $F_{non-flip} \neq 0$, $F_{flip} = 0$ (the " Chew " mechanism[33])

(iii) $F_{non-flip} = 0$, $F_{flip} = 0$ (the " no-compensation " mechanism.[31] The reason for this name is that the $J = 0$ and $J = -1$ amplitudes are linked by a special symmetry, first noted by Gell-Mann.[32] Analogously to the situation in conspiracy theory, when a trajectory passes through $J = 0$ the special symmetry can be satisfied either by having a " compensating " trajectory at $J = -1$ or by extra zeros in the $J = 0$ amplitude—in which case no " compensating " trajectory at $J = -1$ is required).

The analysis is complicated and it is hard to draw firm conclusions, but Chiu et al. obtained the best fits with the third, " no-compensation " behavior. Note that if this is right, A_2 exchange in $\pi^- p \rightarrow \eta n$ should give a similar dip at $\alpha_{A_2} = 0$ (A_2 is a member of the same nonet at P' and therefore might be expected to behave in the same way at $\alpha = 0$). No dip has shown up in the incomplete η production data available so far.

Polarization. Finally there is polarization data.[34] Polarization is an interference effect and is therefore sensitive to such details as the vanishing of the ρ exchange helicity-flip amplitude at $t = -0.6$ BeV^2. It has been fit by Chiu et al.[6] In their fit, the polarization is mainly due to ρ interfering with $(P + P')$ exchange. This choice changes sign between $\pi^+ p$ and $\pi^- p$, and becomes small near $t = -0.6$ BeV^2—two prominent features of the data (Figs. 3 and 4).

Backward πN Scattering. The other region in which simple Regge pole exchange models can be made is backward scattering (small u). The highest known trajectories known are those responsible for the nucleon, (N_α), the 3-3 resonances (Δ_δ), and the 1520 MeV ($I = \frac{1}{2}$) resonance (N_γ).

Data for $d\sigma/du$ are shown in Fig. 5.[35,36] The backward cross section is much bigger for $\pi^+ p \to p\pi^+$ (both $I = \frac{1}{2}$ and $\frac{3}{2}$ exchange) than for $\pi^- p \to p\pi^-$

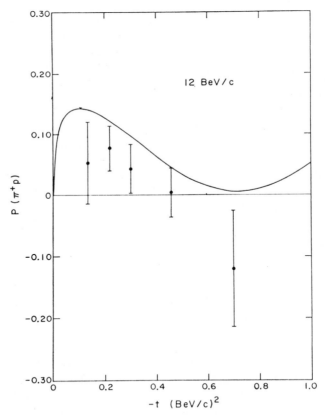

FIG. 3. $\pi^+ p$ polarization data at 12 BeV/c, compared to the $\rho PP'$ exchange model of Rarita et al.[6]

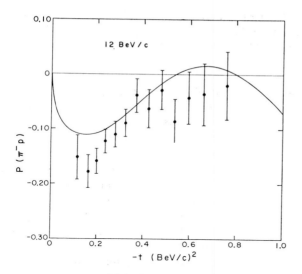

FIG. 4. $\pi^- p$ polarization data at 12 BeV/c, compared to the $\rho P P'$ exchange model of Rarita et al.[6]

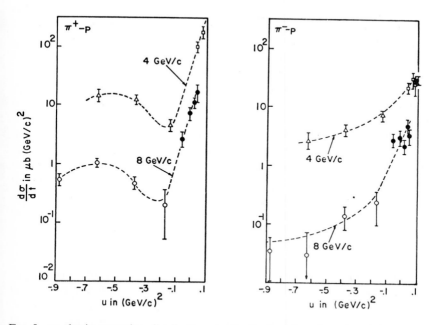

FIG. 5. πp elastic scattering distributions in the backward peak region (taken from Ref. 36).

(pure $I = \frac{3}{2}$ exchange). On this basis, $I = \frac{1}{2}$ exchange is assigned a stronger coupling than Δ_γ exchange. It is also customary to neglect the somewhat lower and more weakly coupled N_δ trajectory in favor of the N_α. The shapes of the two backward cross sections are very different—π^+p sharply peaked, then passing through a dip at $u \approx -0.2$ BeV^2, whereas π^-p is unusually flat. Chiu and Stack[37] have explained this difference as follows. Backward π^-p is dominated by Δ_δ exchange, with $\alpha_\Lambda \sim 0$ as deduced from the energy dependence of $d\sigma/du$. The coupling of the Δ_δ trajectory has no special reason to have a zero in this region (if we ignore Mandelstam-Wang fixed poles, the pole occurring when a Regge α passes through a half-integer value is multiplied by the following factors: a nonsense zero at all unphysical α; i.e., $-\frac{1}{2}$, $-\frac{3}{2}$, $-\frac{5}{2}$, ..., and a wrong signature zero at $...\frac{5}{2}, \frac{1}{2}, -\frac{3}{2}...$ The net effect is a zero at wrong signature nonsense points $-\frac{3}{2}, -\frac{7}{2}, -\frac{11}{2}, ...$). Thus the lack of a dip is not surprising. On the other hand, backward π^+p is dominated by N_α exchange, with $\alpha_{N_\alpha} \sim 0$ to $-\frac{1}{2}$ as deduced from the energy dependence. At $\alpha_{N_\alpha} = -\frac{1}{2}$, a wrong signature nonsense point occurs where the coupling may vanish (the analysis is the same as for Δ_δ except that the wrong signature zeros are now at $...\frac{7}{2}, \frac{3}{2}, -\frac{1}{2}$, $-\frac{5}{2}$, ... and the wrong signature nonsense points at $-\frac{1}{2}$, $-\frac{5}{2}$, $-\frac{9}{2}$, ...). Thus the dip in $d\sigma(\pi^+p)/du$ at $u \approx -0.2$ BeV^2 can be associated with $\alpha_{N_\alpha} = -\frac{1}{2}$. As in the case of ρ exchange, this interpretation of the dip fits in fairly well with α_{N_α} as deduced from the energy dependence of $d\sigma/du$. It also fits spectacularly well with the straight line extrapolation of α_{N_α} from the established $\frac{1}{2}^+$ and $\frac{5}{2}^+$ states and the higher states conjectured to lie on this trajectory by Barger and Cline[38] (Fig. 6).

As at $\alpha_\rho = 0$, the dip could have been spoiled by a Mandelstam-Wang, Jones-Teplitz fixed pole, so once again the fixed pole apparently has a small coefficient.

Another interesting feature of backward πN scattering is that for each trajectory at $u = 0$, there is another trajectory with the same α and opposite parity. This is a consequence of the MacDowell symmetry, and is an example of "conspiracy." At present, we have no direct experimental evidence for the parity doubling of trajectories at $u = 0$, but Stack[39] has pointed out that —unlike a single trajectory model—a model with a parity doubled trajectory leads to a large polarization in nearly-backward scattering. Thus polarization measurements in this region would be very interesting.

Intermediate Energies. At intermediate energies, above the region covered by detailed phase shift analysis, there are still many peaks and dips reflecting the presence of distinguishable direct channel resonances. These peaks and dips are especially clear in the careful measurements made at $0°$ and $180°$. Their spins and parities are not firmly established in most cases.

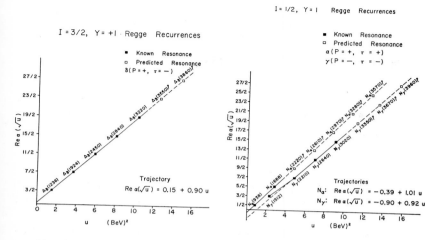

FIG. 6. Plot of the $Y = +1$ fermion Regge recurrence, taken from Barger and Cline.[38] Note that the spins of the higher "known resonances" on the plot have not been verified experimentally.

Barger and Cline[38] have made the striking hypothesis that all these resonances lie in straight line extrapolations of the three leading nucleon trajectories N_α, Δ_δ, and N_γ (Fig.. 6), and have shown that this hypothesis is consistent with the existing data. The straight lines are defined by the lowest two or three states on each trajectory, whose spin-parity assignments are already established.

The successful checks of the hypothesis thus far include:

(i) each visible resonance occurs at a mass consistent with an expected physical state on one of the three straight lines

(ii) the parity of a resonance determines the sign of its contribution to the 180° cross section and for a given background this sign determines whether the resonance appears as a peak (constructive interference) or dip (destructive interference). Assuming a slowly varying background, Barger and Cline show that the observed dip-peak sequence[40] is consistent with the parities assigned in Fig. 6

Further checks which must be carried out to really establish the hypothesis are:

(iii) the spins of the higher resonances in Fig. 6 must be determined directly

(iv) the elasticity of the higher resonances is predicted to be very low (since Barger and Cline are assigning a high statistical weight $(2J + 1)$ to small peaks), and this has to be checked.

If the trajectories are really straight lines, then the MacDowell symmetry implies that the resonances should have doubles of the opposite parity. This is clearly not the case for the nucleon and 3-3 resonance. Barger and Cline[41] have shown, however, that some of the other parity doublets do appear among Lovelace's nine new resonances. The overall status of the parity doublets is still unclear.

A curious feature of the Barger-Cline proposal is that it has the spin of the trajectories rising as $\alpha = J \sim s$, whereas the orbital angular momentum for the πN channel with fixed-range forces rises only like

$$l = |r \times p_{CM}| \sim s^{1/2}.$$

This implies that the high spin resonances would decay principally into secondaries with sufficiently high spin (e.g., pion recurrence plus nucleon recurrence) to keep the orbital angular momentum low.

This novel idea is at least qualitatively consistent with the low partial widths Barger and Cline assign for decay into the low spin πN channel, and with the relatively narrow total widths of these resonances (the idea being that in spite of the many channels open, all but the relatively few channels containing high spin secondaries are blocked by high centrifugal barriers).

Interference Model. In their detailed fitting of cross sections, Barger and Cline (as well as many others) make a specific assumption about the background term which is to be added to the direct channel resonances. Namely, they approximate it by the leading Regge pole exchanges:

$$F = F_{res} + F_{Regge} \tag{3}$$

Now the sum over all partial waves in the direct channel adds up to the correct full amplitude, and the sum over dominant direct channel resonances is an approximation to this.[42] Similarly, the sum over all partial waves in the crossed channel, or equivalently, over all terms in the Sommerfeld-Watson representation for the crossed channel, adds up to the *same* full amplitude, and the sum over dominant exchanged Regge poles is an approximation to this. Evidently, then, if we pushed (3) to the limit by including more and more terms, the sum over all direct channel contributions plus all exchange terms would eventually add up to *twice* the full amplitude. Thus, care must be exercised in applying Eq. (3).

New light has been shed on the validity of Eq. (3) by the FESR analysis of Dolen et al.[5] The FESR

$$S_0 = \frac{1}{N} \int_0^N \text{Im}\, F(v)\, dv = \sum_i \frac{\beta_i N^{\alpha_i}}{(\alpha_i + 1)\Gamma(\alpha_i + 1)} \tag{4}$$

implies that the Regge terms at any energy (right side) are given by the average of the resonance terms up to that energy (left side). Thus Dolen et al. argue that one should take

$$F = F_{res} - \bar{F}_{res} + F_{Regge}; \qquad (5)$$

double counting is committed if the *average* of the resonance terms is not subtracted out. The usual interference model (Eq. (3)) holds only when $\bar{F}_{res}$ happens to be very small.

Finally it is important to note that this observation, while casting doubt on some specific applications of the interference *model* (Eq. (3)), does not by any means invalidate the Barger-Cline hypothesis or the importance of interference *effects*. The point is only that these will be understood more reliably if double counting is avoided.

References

1. C. E. Jones and V. L. Teplitz, *Phys. Rev.*, **159**, 1271 (1967).
2. S. Mandelstam and L. L. Wang, *Phys. Rev.*, **160**, 1490 (1967).
3. A. Logunov, L. D. Soloviev, and A. N. Tavkhelidze, *Phys. Letters*, **24B**, 181 (1967).
4. K. Igi and S. Matsuda, *Phys. Rev. Letters*, **18**, 625 (1967).
5. R. Dolen, D. Horn, and C. Schmid, *Phys. Rev. Letters*, **19**, 402 (1967), and California Institute of Technology report CALT-68-127 (1967).
6. W. Rarita, R. J. Riddell, C. B. Chiu, and R. J. N. Phillips, *Phys. Rev.*, **165**, 1615 (1968).
7. R. Logan, *Phys. Rev. Letters*, **14**, 414 (1965).
8. R. Phillips and W. Rarita, *Phys. Rev.*, **139**, B1336 (1965).
9. G. Höhler, J. Baacke, H. Schlaile, and P. Sonderegger, *Phys. Letters*, **20**, 79 (1966).
10. F. Arbab and C. B. Chiu, *Phys. Rev.*, **147**, 1045 (1966).
11. I. Mannelli et al., *Phys. Rev. Letters*, **14**, 408 (1965).
12. P. Sonderegger et al., *Phys. Rev. Letters*, **14**, 763 (1965), and *Phys. Letters*, **20**, 75 (1966).
13. P. Bonamy et al., *Phys. Letters*, **23**, 501 (1966).
14. H. Högaasen and W. Fischer, *Phys. Letters*, **22**, 516 (1966).
15. R. K. Logan, J. Beaupre, and L. Sertorio, *Phys. Rev. Letters*, **18**, 259 (1967).
16. W. Rarita and B. Schwarzschild, *Phys. Rev.*, **162**, 1378 (1967).
17. L. Sertorio and M. Toller, *Phys. Rev. Letters*, **19**, 1146 (1967).
18. V. M. de Lany, D. J. Gross, I. J. Muzinich, and V. L. Teplitz, *Phys. Rev. Letters*, **18**, 148 (1967).
19. C. Chiu and J. Finkelstein, UCRL preprint 17318 (1967).
20. R. Phillips, *Nuovo Cimento*, **45**, 245 (1966).
21. R. Logan and L. Sertorio, *Phys. Rev. Letters*, **17**, 834 (1966).
22. G. Altarelli, A. Borgese, F. Buccella, and M. Colocci, *Nuovo Cimento*, **48A**, 245 (1967).
23. B. Desai, D. Gregorich, and R. Ramachandran, *Phys. Rev. Letters*, **18**, 565 (1967).
24. K. Igi, *Phys. Rev. Letters*, **9**, 76 (1962).

25. Some variation in these parameters is certainly possible, however, as emphasized by N. Cabibbo, L. Horwitz, J. J. Kokkedee, and Y. Ne'eman, *Nuovo Cimento*, **45A**, 275 (1966).
26. K. J. Foley, S. J. Lindenbaum, W. A. Love, S. Ozaki, J. J. Russell, and L. C. L. Yuan, *Phys. Rev. Letters*, **11**, 425, 503 (1963); **15**, 45 (1965).
27. V. Barger and L. Durand, III, *Phys. Rev. Letters*, **19**, 1295 (1967).
28. C. T. Coffin et al., *Phys. Rev. Letters*, **15**, 838 (1965); **17**, 458 (1966).
29. J. Orear et al., *Phys. Rev.*, **152**, 1162 (1966).
30. S. Frautschi, *Phys. Rev. Letters*, **17**, 722 (1966).
31. C. B. Chiu, S. Y. Chu, and L. L. Wang, *Phys. Rev.*, **161**, 1563 (1967).
32. M. Gell-Mann, *Proc. 1962 Intern. Conf. High Energy Physics, CERN*, p. 539.
33. G. F. Chew, *Phys. Rev. Letters*, **16**, 60 (1966).
34. M. Borghini et al., *Physics Letters*, **24B**, 77 (1967).
35. H. Brody et al., *Phys. Rev. Letters*, **16**, 828 (1966).
36. A. Ashmore et al., *Phys. Rev. Letters*, **19**, 460 (1967).
37. C. B. Chiu and J. Stack, *Phys. Rev.*, **153**, 1575 (1967).
38. V. Barger and D. Cline, *Phys. Rev. Letters*, **16**, 913 (1966); *Phys. Rev.* **155**, 1792 (1967).
39. J. Stack, *Phys. Rev. Letters*, **16**, 286 (1966).
40. S. W. Kormanyos, A. D. Krisch, J. R. O'Fallon, K. Ruddick, and L. G. Ratner, *Phys. Rev. Letters*, **16**, 709 (1966).
41. V. Barger and D. Cline, *Phys. Rev. Letters*, **20**, 298 (1968).
42. An example in the case of backward πN scattering is F. N. Dikmen, *Phys. Rev. Letters*, **18**, 798 (1967).

Current Algebra Determination of Low-Energy πN Parameters*

HOWARD J. SCHNITZER

*Department of Physics, Brandeis University,
Waltham, Massachusetts*

The original proposal of Gell-Mann[1] to place the axial-vector current and vector current on equal footing by means of local equal-time commutation rules supplemented by the partial conservation of the axial vector current (PCAC) has proved to be extremely fruitful in adding to our understanding of particle phenomena. The first dramatic success of this program was achieved by Adler and Weisberger[2] (AW), who derived a sum rule which correlated the axial-vector coupling constant with an integral over πN total cross sections. It was subsequently recognized that the AW calculation could be recast as a low energy theorem for the s-wave scattering lengths[3] which does not make assumptions about unsubtracted dispersion relations as required in the AW formulation. A particular advantage of the Weinberg presentation[4] of the theorem is that it clearly exhibits the assumptions made in the derivation, while providing a framework for the extension of the current-algebra method to the prediction of other low energy πN parameters. This is the point of view we adopt here.

It is generally assumed that the vector and axial-vector current satisfy the equal-time commutation rules of a local chiral $SU(2) \times SU(2)$ algebra. The axial-vector current $A_\mu^a(x)$ with isospin a has the commutation rule

$$[A_0^a(x), A_\mu^b(y)]\delta(x_0 - y_0) = 2i\varepsilon_{abc} \, \delta^4(x - y)V_\mu^c(x) + \text{ST} \qquad (1)$$

and

$$\delta(x_0 - y_0)[A_0^a(x), \partial^\mu A_\mu^b(y)] = c\sigma_{ab}(x)\delta^4(x - y) \qquad (2)$$

where $V_\mu^c(x)$ is the vector current, $\sigma_{ab}(x) = \sigma_{ba}(x)$ is a scalar field, and ST indicates possible Schwinger terms.[5] The current commutation relations and PCAC provide a connection between off-shell πN and weak

---

* Supported in part by the National Science Foundation.

axial-vector nucleon scattering by means of the basic relation

$$\int d^4x \, d^4y e^{iq\cdot x} e^{-ik\cdot y} \langle p_2 | T\{\partial^\mu A_\mu^{\,b}(x), \partial^\nu A_\nu^{\,b}(y)\} | p_1 \rangle$$

$$= q^\mu k^\nu \int d^4x \, d^4y e^{iq\cdot x} e^{-ik\cdot y} \langle p_2 | T\{A_\mu^{\,b}(x) A_\nu^{\,a}(y)\} | p_1 \rangle$$

$$- 2q^\mu \varepsilon_{abc} \int d^4x e^{i(q-k)\cdot x} \langle p_2 | V_\mu^{\,c}(x) | p_1 \rangle \tag{3}$$

$$- c \int d^4x e^{i(q-k)\cdot x} \langle p_2 | \sigma_{ab}(x) | p_1 \rangle$$

where $k^\mu(q^\mu)$ and $p_1(p_2)$ are the initial (final) pion momenta and nucleon momenta, respectively. One *defines* the field as

$$\phi^a(x) = \frac{1}{im_\pi F_\pi} \partial^\mu A_\mu^{\,a}(x) \tag{4}$$

which relates the off-mass-shell πN amplitude to the basic identity (3) means of the usual reduction formula

$$T^{ba}(p_2, q; p_1, k) = i\left(\frac{E_2}{m}\frac{E_1}{m}\right)^{1/2}(2\pi)^3 \frac{(q^2 - m_\pi^2)(k^2 - m_\pi^2)}{m_\pi^4 F_\pi^2}$$

$$\times \int d^4x e^{iq\cdot x} \langle p_2 | T\{\partial^\mu A_\mu^{\,b}(x) \, \partial^\nu A_\nu^{\,a}(0)\} | p_1 \rangle \tag{5}$$

The pion decay constant F_π can be estimated by the Goldberger-Treiman relation

$$F_\pi = \frac{2mg_A}{g_{\pi N}(0)} \simeq \frac{2mg_A}{g_{\pi N}} \tag{6}$$

which is derived from the usual arguments of PCAC. It is the *assumption* of PCAC that (4) leads to continuations of matrix elements in the pion mass q^2 which have the *normal* variation expected from the dynamical singularities encountered in the continuation.[6] This then suggests that one should expect an error of roughly $10-15\% \simeq (m_\pi/3m_\pi)^2$ in calculations depending on PCAC. The off-shell amplitude can be separated into invariant amplitudes and partial waves exactly as the physical amplitude. Thus

$$T^{ba}(s, t, q^2, k^2) = A^{ba}(s, t, q^2, k^2) - \gamma \cdot Q B^{ba}(s, t, q^2, k^2) \tag{7}$$

with

$$T^{ba} = \delta_{ab} T^{(+)} + \tfrac{1}{2}[\tau_b, \tau_a] T^{(-)}$$

$$W^2 = s = (p_1 + k)^2, \, t = (p_1 - p_2)^2, \text{ and } u = (p_2 - k)^2.$$

The partial wave amplitudes in the center of mass system are given by the standard decompositions (with $q^2 = k^2$ for convenience).

$$f(s, \cos\theta, q^2) = \frac{m}{4\pi W} T(s, t, q^2) \tag{8a}$$

$$|f|^2 = \frac{d\sigma}{d\Omega}$$

$$f = f_1 + \frac{(\boldsymbol{\sigma} \cdot \mathbf{q})(\boldsymbol{\sigma} \cdot \mathbf{k})}{|\mathbf{q}||\mathbf{k}|} f_2 \tag{8b}$$

$$f_{l\pm}(s, q^2) = e^{i\delta_{l\pm}} \sin\delta_{l\pm}/|q|^2 \tag{8c}$$

$$= \frac{1}{2} \int_{-1}^{1} dx[f_1(x)P_l(x) + f_2(x)P_{l\pm 1}(x)],$$

$$= \frac{1}{16\pi W^2} \{[(W+m)^2 - q^2][A_l - (W - m)B_l] \tag{8d}$$

$$- [(W-m)^2 - q^2][A_{l\pm 1} + (W+m)B_{l\pm 1}]\}$$

$$A_l(s, q^2) = \frac{1}{2} \int_{-1}^{1} dx P_l(x)A(s, t, q^2) \tag{8e}$$

$$|q|^2 = [(W+m)^2 - q^2][(W-m)^2 - q^2]/4W^2 \tag{8f}$$

It is important to note that the partial-wave amplitudes $f_{l\pm}(s, q^2)$ have kinematical singularities in q^2 and hence are not the appropriate objects for the application of PCAC. Rather it is the invariant amplitudes $A(s, t, q^2)$ and $B(s, t, q^2)$ which are expected to have only dynamical singularities in q^2, and thus are expected to satisfy the above definition of PCAC.

The one nucleon contribution to (7) follows immediately from the single nucleon matrix elements of the axial-vector current:

$$A_N^{(-)}(s, t, q^2) = 0 \tag{9a}$$

$$A_N^{(+)} = 4\left(\frac{q^2 - m_\pi^2}{m_\pi^2 F_\pi}\right)^2 [mg_A(q^2) - q^2 h_A(q^2)]g_A(q^2) \tag{9b}$$

$$B_N^{(-)} = \frac{g_{\pi N}(q^2)}{s - m^2 + i\varepsilon} + \frac{g_{\pi N}(q)^2}{u - m^2 + i\varepsilon} + 2\left(\frac{q^2 - m_\pi^2}{m_\pi^2 F_\pi}\right)^2 g_A^2(q^2) \tag{9c}$$

and

$$B_N^{(+)} = \frac{g_{\pi N}(q^2)}{s - m^2 + i\varepsilon} - \frac{g_{\pi N}(q^2)}{u - m^2 + i\varepsilon} \tag{9d}$$

where

$$\langle p_2 | A_\mu^b(0) | p_1 \rangle = \frac{1}{(2\pi)^3} \left(\frac{m}{E_2} \frac{m}{E_1} \right)^{1/2} \bar{u}(p_2)[g_A(q^2)i\gamma_\mu\gamma_5 + h_A(q^2)iq_\mu\gamma_5]\tau_b u(p_1)$$

(10)

and

$$(q^2 - m_\pi^2)[-2mg_A(q^2) + q^2h_A(q^2)] = F_\pi m_\pi^2 g_{\pi N}(q^2)$$

(11)

One notices that (9) is essentially the Born approximation in gradient coupling $ps(pv)$ theory. The additional nonsingular terms in (9b) and (9c). compared to the usual $ps(ps)$ Born terms are essential for the correct prediction of the s-wave scattering lengths. The commutator terms are also expressed in terms of the invariants. The contribution of the vector current is

$$A_c^{(-)}(s, t, q^2) = -2\left(\frac{q^2 - m_\pi^2}{m_\pi^2 F_\pi} \right)^2 \left(s + \frac{1}{2}t - m^2 - q^2 \right) \frac{F_2(t)}{m}$$

(12a)

$$A_c^{(+)} = 0$$

(12b)

$$B_c^{(-)} = -4\left(\frac{q^2 - m_\pi^2}{m_\pi^2 F_\pi} \right)^2 [F_1(t) + F_2(t)]$$

(12c)

and

$$B_c^{(+)} = 0$$

(12d)

where $F_{1,2}(t)$ are the Dirac isovector electromagnetic form factors normalized to $F_1(0) = \frac{1}{2}$ and $F_2(0) = 1.85$. The σ-commutator only involves a single invariant amplitude which one can write as

$$A_s^{(+)} = C\left(\frac{q^2 - m_\pi^2}{m_\pi^2 F_\pi} \right)^2 G(t)$$

(13)

where C is the constant that appears in Eq. (2), and $G(t)$ is defined in the obvious way from (3) and (5).

We now have enough information to discuss the current algebra predictions for the low energy phenomena in a coherent way. The basic result is the correct prediction of the s-wave scattering lengths, which is easily obtained once one realizes that aside from pole terms, which must be discussed separately, the weak amplitude may be neglected at threshold. This is argued as a direct consequence of PCAC, which postulates that this term is as smooth as is possible to expect from dynamical singularities alone. Thus once one has separated out the poles, the coefficients of $q^\mu k^\nu$ should be of order $g_{\pi N}^2/m_i^2$ at threshold, where m_i is some large internal mass. Hence the terms quadratic in momenta (with poles removed) are of order $g_{\pi N}{}^2 m_\pi{}^2/m_i{}^2$ at threshold. On the other hand, the term linear in momentum in (3) is of order $g_{\pi N}{}^2 m_\pi/m$, so that the quadratic terms may be neglected *at threshold* compared to the linear term. This particular consequence of PCAC can be restated in terms of the invariant amplitudes. Let $v = p_i \cdot k/m$ and v_t its value at threshold. Since the forward amplitude is proportional to $A(v, q^2) - vB(v, q^2)$, the S-wave scattering lengths are proportional to this combination with $v \to v_t$. In this language PCAC requires that in the forward direction

$$\lim_{v \to v_t} \{[A(v) - vB(v)]_{\text{weak}} - \text{poles}\}$$

be of order

$$g_{\pi N}{}^2 \frac{v_t^2}{m^2} \sim g_{\pi N}{}^2 \frac{q^2}{m^2}$$

while the linear term be of order $g_{\pi N} v_{t/m}$ and hence dominant.

To continue the argument one must examine the poles coming from the low-lying intermediate states in the weak amplitude. For the nucleon pole one applies Eq. (11) together with (10) to show that

$$\lim_{v \to v_t} [A_N{}^{(\pm)} - vB_N{}^{(\pm)}] \sim g_{\pi N}{}^2 \frac{q^2}{m^2} \tag{14}$$

which demonstrates that the nucleon term is also negligible at threshold compared to the linear term. It is possible to separate our other low-lying states from the weak amplitude such as the N^* or Roper resonance and compute their contributions to the scattering lengths. Explicit calculation shows that these are also of order $g_{\pi N} v_t{}^2$. However, if there were a low-lying $S_{1/2}$ baryon pole, it would give rise to a nonneglible contribution to the threshold amplitude of order

$$g_{\pi N}{}^2(q^2/M^2 - m^2 + 2mv_t)$$

where M is the hypothetical resonance mass. Clearly the neglect of the weak amplitude in the scattering length calculation depends on the absence of such states.

Finally it is argued that the σ-term is also negligible. This uses the Adler consistency condition[7] which states the scattering amplitude (5) vanishes if $k^2 \to m_\pi{}^2$ and $q^\mu \to 0$. This is easily verified by combining (3) with (5) and noting that only the weak amplitude has a pion pole, but it does not have a pole at $q^\mu \to 0$. The implication of this constraint is that the σ-term and isotopic even part of the weak amplitude are of the same order of magnitude at this point, which means that the constant C in (13) is of order $g_{\pi N}{}^2 m_\pi{}^2/m^2$, so that the σ-term is also negligible. This argument could fail if there were important low-lying $T = 0, J = 0$ π-π enhancements in the t-channel, in which case the extrapolation from the Adler consistency point to the physical threshold would not be slowly varying. The success of the extrapolation is a strong argument for the absence of such enhancements.

Putting the arguments together one sees that only the isovector commutator is relevant for the calculation of the S-wave scattering lengths. Thus from equations (8) and (12) one predicts[3]

$$a_{1/2} = \frac{2g_{\pi N}{}^2 m_\pi}{8\pi m^2 g_A{}^2} \frac{1}{1 + m_\pi/m} \simeq 0.20 \frac{1}{m_\pi} \tag{15}$$

and

$$a_{3/2} = -\tfrac{1}{2}a_{1/2} \simeq -0.10 \frac{1}{m_\pi}$$

which are in good agreement with the experimental values of $a_{1/2} = 0.171 \pm 0.005$ $1/m_\pi$ and $a_{3/2} = -0.088 \pm 0.004$ $1/m_\pi$ (within the 15% error expected of PCAC). Since the actual value of $a_{1/2} + 2a_{3/2}$ is quite small, one has experimental verification of the Adler consistency condition. The advantage of this formulation of the current algebra predictions is that it makes no assumptions about dispersion relations and the number of subtractions. If one does assume that the isotopic odd part of the πN scattering amplitude satisfies an unsubtracted dispersion, then one has the Goldberger-Miyazawa-Oehme sum rule. If this sum rule is combined with (15) and the limit $m_\pi \to 0$ is taken, then the Adler-Weisberger sum rule[2,3]

$$\frac{1}{g_A{}^2} = 1 - \frac{2m^2}{g_{\pi N}{}^2} \frac{1}{\pi} \int_{s_0}^{\infty} ds \frac{\sigma_+(s) - \sigma_-(s)}{s - m^2} \tag{16}$$

is obtained

An interesting observation has been made by Sakurai,[8] who connects the scattering lengths with the ρ-dominance assumption. He notes that the ad hoc assumptions that the threshold S-wave scattering can be described entirely in terms of ρ meson exchange, evaluated in Born approximation, also gives a reasonable estimate of the scattering lengths. If this estimate is equated to the current algebra prediction one finds

$$F_\pi{}^2 = \frac{g_\rho{}^2}{2m_\rho{}^2} \tag{17}$$

where g_ρ is defined through the matrix element

$$\langle 0 | V_\mu{}^a(0) | \rho, p; b \rangle = \frac{\delta_{ab} g_\rho \varepsilon_\mu{}^\rho(p)}{(2\pi)^3 (2\omega_\rho)^{1/2}}$$

In obtaining (17), known as the Kawarabayaski-Suzuki-Fayazzµdin-Riazzudin (FSFR)[9] relation, use is made of the Goldberger-Treiman relation, ρ-dominance, and the assumption of the universality of the isovector current. It appears to agree well with experiment, and is also useful in other applications of current algebra.[10] However *all* known " derivations " from current algebra must be supplemented by additional dynamical assumptions; there is no derivation from current algebra and ρ-dominance alone.

It is possible to predict other features of low-energy π-N scattering if detailed information on the weak amplitude is available. For example, the part of the scattering amplitude which vanishes at threshold as $|\mathbf{q}|^2$ faster than the scattering length leads to a determination of the p-wave scattering lengths and S-wave effective ranges. One no longer can apply *soft* pion techniques ($q^\mu \to 0$) since the terms of interest involve nonzero pion kinetic energy, however PCAC and the off-shell continuation in q^2 may still be used to advantage. From Eq. (8), keeping terms to the required order in $|\mathbf{q}|^2$, can easily show that [11]

$$\frac{f_{1-}}{|\mathbf{q}|^2} \xrightarrow[\text{threshold}]{} \frac{1}{4\pi(m+v_t)} \left\{ \frac{2m}{3} \left[\frac{\partial A}{\partial t} - v_t \frac{\partial B}{\partial t} \right] - \frac{1}{4m} [A + (2m+v_t)B] \right\}_{t=0}$$

(18a)

$$\frac{f_{1+}}{|\mathbf{q}|^2} \to \frac{1}{4\pi(m+v_t)} \frac{2m}{3} \left(\frac{\partial A}{\partial t} - v_t \frac{\partial B}{\partial t} \right)_{t=0} \qquad (18b)$$

and

$$\frac{\partial f_{0+}}{\partial |\mathbf{q}|^2} \to \frac{1}{4\pi} \left\{ \frac{(m+v_t)}{v_t} \left(\frac{\partial A}{\partial s} - v_t \frac{\partial B}{\partial s} \right) \right.$$
$$\left. - 2\left(\frac{\partial A}{\partial t} - v_t \frac{\partial B}{\partial t} \right) - \frac{1}{2v_t(m+v_t)} (A + mB) \right\}_{t=0} \quad (18c)$$

Since A and B separately are *not* negligible at threshold, it is obvious that the weak amplitude cannot be neglected in computing these quantities, although the omission of the σ-term is still justified. Hence a model of the weak amplitude is required, which by reasons of unitarity means one needs the amplitude for the axial-vector production of hadrons from nucleons. Since it is possible to show that PCAC that the weak axial production of hadron states with J $3/2$ $-$ and $J \geq 5/2$ give negligible contributions to Eq. (18), the task of model making is considerably simplified. The nucleon pole and N^* production are expected to be dominant, although the weak production of s- and p-wave continua states and Roper resonance might also be important. A particularly simple model involving direct production of a narrow N^* state has been investigated[11,12] which proved to be reasonably successful.

One finds that the nucleon contribution to the p-wave scattering lengths, [computed from (9), (18a), and (18b)] reproduces the usual result from Born approximation as required by the equivalence of $ps(ps)$ and $ps(pv)$ theories at the nucleon pole. Furthermore, because the effects of the continuation off-shell of $A_{N^*}(s, t, q^2)$ and $B_{N^*}(s, t, q^2)$ are not very important in the narrow N^* resonance model, the N^* contribution is essentially the same as that computed by Amati and Fubini[13] using an isobar model of πN scattering. We indicate the predictions of our model schematically as follows[11]:

$$\frac{f_{1+}{}^{(-)}}{|\mathbf{q}|^2} \xrightarrow[\text{threshold}]{} f^2 \left[-\frac{2}{3} + N^* - \frac{2m_\pi}{3mg_A{}^2} F_2(0) \right] \frac{1}{m_\pi{}^3} \qquad (19a)$$

$$\frac{f_{1-}{}^{(-)}}{|\mathbf{q}|^2} \rightarrow f^2 \left[-\frac{2}{3} + N^* - \frac{2m_\pi}{3mg_A{}^2} F_2(0) + \frac{2m_\pi}{mg_A{}^2} (F_1(0) + F_2(0)) \right] \frac{1}{m_\pi{}^3}$$

$$\qquad (19b)$$

$$\frac{f_{1+}{}^{(+)}}{|\mathbf{q}|^2} \rightarrow f^2 \left(+\frac{2}{3} + N^* \right) \frac{1}{m_\pi{}^3} \qquad (19c)$$

and

$$\frac{f_{1-}{}^{(+)}}{|\mathbf{q}|^2} \rightarrow f^2 \left(-\frac{4}{3} + N^* \right) \frac{1}{m_\pi{}^3} \qquad (19d)$$

where $f^2 = (g_{\pi N}{}^2/4\pi)(m_\pi/2m)^2 \simeq 0.08$. In deriving such expressions, terms of higher order in m_π/m are dropped as usual; however, some care must be exercised since terms of order $q^2/M^{*2} - m^2$ are encountered, which must be kept. Typical numerical values are shown in Table I.[11] Evidently the

TABLE I

Contributions of the Various Terms to the p-Wave Scattering Lengths.

Amplitude	N	N^*	Commutator	Total	Experiment	
					HW	Roper
$f_{1+}^{(-)}/|\mathbf{q}|^2$	-0.054	-0.018	-0.007	-0.079	-0.081	-0.081
$f_{1-}^{(-)}/|\mathbf{q}|^2$	-0.054	$+0.019$	$+0.033$	-0.002	-0.021	-0.001
$f_{1+}^{(+)}/|\mathbf{q}|^2$	$+0.054$	$+0.074$	0	$+0.128$	$+0.134$	$+0.134$
$f_{1-}^{(+)}/|\mathbf{q}|^2$	-0.108	$+0.047$	0	-0.061	-0.059	-0.039

[a] The theoretical results are compared to the experimental analyses of Hamilton and Woolcock and Roper et al.[14]

predictions agree with the experimental scattering lengths within the limits of accuracy expected of PCAC.

The S-wave effective ranges have been computed from the same model, the results which we summarize as follows:[11]

$$\mathrm{Re}\,\frac{\partial f_{0+}^{\,(+)}}{\partial|\mathbf{q}|^2}\xrightarrow[\text{threshold}]{} f^2(N^*)\,\frac{1}{m_\pi^3}\simeq -0.072\,\frac{1}{m_\pi^3}\qquad (20a)$$

and

$$\mathrm{Re}\,\frac{\partial f_{0+}^{\,(-)}}{\partial|\mathbf{q}|^2}\to f^2\!\left(2+N^*+\frac{2}{g^2}\left\{F_1(0)-\frac{m_\pi}{m}\left[F_1(0)-4m_\pi^{\,2}\frac{\partial F_1(0)}{\partial t}\right]\right\}\right)$$

$$\times\frac{1}{m_\pi^3}\approx +0.25\frac{1}{m_\pi^3}\qquad (20b)$$

These predictions can be compared with the analysis of Hamilton and Woolcock[14]

$$\mathrm{Re}\,\frac{\partial f_{0+}^{\,(+)}}{\partial|\mathbf{q}|^2}\to -0.042\,\frac{1}{m_\pi^3}$$

and

$$\mathrm{Re}\,\frac{\partial f_{0+}^{\,(-)}}{\partial|\mathbf{q}|^2}\to 0.010\,\frac{1}{m_\pi^3}\qquad (21)$$

We suspect that the poor agreement between (20b) and the experimental analysis may be due to the neglect of the t-channel σ and ρ poles in the weak amplitude. The low energy continuum also deserves more attention than is given in this simple model.

It was argued that the non-spin-flip Adler-Weisberger relation was equivalent to the current algebra of S-wave scattering lengths. It is not difficult to see that the calculation of the p-wave scattering lengths involves some knowledge of the forward, spin-flip amplitude. However, since we have shown that one has to introduce a model for the weak axial-vector-nucleon scattering to do so, this strongly suggested that no (model independent) spin-flip sum rule of the AW type should exist. A spin-flip sum rule has been derived from current algebra and PCAC which depends on information from the weak amplitude[15]; the crucial point in the derivation being the choice of a set of tensor covariants free of kinematical singularities. Although only qualitative checks have been made, this sum rule seems to be satisfied.

Another development in current algebra with applications to low-energy pion scattering is the method of effective Lagrangians.[16] The effective or phenomenological Lagrangian which incorporates current algebra and PCAC is a device for reproducing the usual results of current algebra with much less labor. Although the language of perturbative Lagrangian field theory is used, this machinery is used only to the extent of constructing smooth off-shell amplitudes which satisfy the constraints of current algebra and PCAC. As developed to date it is not the basis for constructing a complete field theory. The original work of Weinberg showed that for soft pions the effective Lagrangian could be written

$$\mathscr{L}_{\text{eff}} = -\overline{N}\left[\gamma^\mu \, \partial_\mu + m + \frac{g_{\pi N}}{2m} \, i\gamma^\mu\gamma_5 \, \tau \cdot \partial_\mu \pi \right.$$

$$\left. + \frac{g_{\pi N}^{\,2}}{4m^2}\left(\frac{1}{g_A}\right)^2 i\gamma^\mu \tau \cdot \pi \times \partial_\mu \pi + \cdots \right]N \quad (22)$$

where the πN scattering lengths are given by the $\pi \times \partial_\mu \pi$ term. The method can be extended to non-soft pions and applied to the p-wave scattering lengths and S-wave effective ranges.[17]

We have seen that the model independent results of current algebra, the s-wave scattering lengths and Adler-Weisberger relation, are in spectacular agreement with experiment. Furthermore, the model dependent calculations give sufficiently good results to justify the methods employed. It is clear that further improvements are linked to a better understanding of the weak amplitude. The most ambitious study to date, by Adler,[12] exemplifies the kind of calculation required for the next generation of results in this area.

References

1. M. Gell-Mann, *Physics*, **1**, 63 (1964).
2. S. L. Adler, *Phys. Rev. Letters*, **14**, 1051 (1965), *Phys, Rev.*, **140**, B736 (1965); W. I. Weisberger, *Phys. Rev. Letters*, **14**, 1047 (1965), *Phys. Rev.*, **143**, 1302 (1966).
3. Y. Tomozawa, *Nuovo Cimento*, **46A**, 707 (1967); A. P. Balachandran, M. G. Gundzik, and F. Nicodemi, *Nuovo Cimento*, **44A**, 1257 (1966); K. Raman and E. C. G. Sudarshan, *Phys. Rev. Letters*, **21**, 450 (1966); B. Hambrecht, Cambridge University report (unpublished); and S. Weinberg, *Phys. Rev. Letters*, **17**, 616 (1967).
4. S. Weinberg, Ref. 3.
5. If we interpret the T-product in our reduction formula as a current correlation function then the Schwinger terms play no role in our discussion. L. S. Brown, *Phys. Rev.*, **150**, 1138 (1966); R. P. Feynman, unpublished.
6. If we had chosen a different definition for the pion field, $\phi(x) = C\Box^{20} \, \partial^\mu A_\mu(x)$ say, then one would not expect this to be true. See, e.g., S. Coleman, *Proc. Erice Summer School*, 1967, to be published. Among the basic references on PCAC are Y. Nambu, *Phys. Rev. Letters*, **4**, 380 (1960); M. Gell-Mann and M. Levy, *Nuovo Cimento*, **16**, 705 (1960).

7. S. L. Adler, *Phys. Rev.*, **137**, B1022 (1965), *Phys. Rev.*, **139**, B1638 (1965).
8. J. J. Sakurai, *Phys. Rev. Letters*, **17**, 552 (1966).
9. K. Kawarabayashi and M. Suzuki, *Phys. Rev. Letters*, **16**, 225 (1966); Fayazuddin and Riazuddin, *Phys, Rev.*, **147**, 1071 (1966); F. J. Gilman and H. J. Schnitzer, *Phys. Rev.*, **150**, 1362 (1966); J. J. Sakurai, Ref. 8; M. Ademollo, *Nuovo Cimento*, **46**, 156 (1966).
10. S. Weinberg, *Phys. Rev. Letters*, **18**, 507 (1967).
11. H. J. Schnitzer, *Phys. Rev.*, **158**, 1471 (1967).
12. K. Raman, *Phys. Rev. Letters.*, **17**, 983 (1966), **18**, 432E (1967), *Phys. Rev.*, **159**, 1501 (1967); K. Ishida and A. Takahashi, Yamagata University preprint, to be published. More detailed models have been studied by S. L. Adler, *Ann. Phys. (N.Y.)*, **50**, 189 (1968).
13. D. Amati and S. Fubini, *Ann. Rev. Nucl. Sci.*, **12**, 359 (1967).
14. J. Hamilton and W. S. Woolcock, *Rev. Mod. Phys.*, **35**, 737 (1963); L. D. Roper, R. M. Wright, and B. T. Feld, *Phys. Rev.*, **138**, B190 (1965).
15. I. S. Gerstein, *Phys. Rev.*, **161**, 1631 (1967); H. Goldberg and F. Gross, *Phys. Rev.*, **162**, 1350 (1967); L. Maiani and G. Preparata, *Nuovo Cimento*, **49B**, 237 (1967); S. L. Adler, unpublished.
16. S. Weinberg, *Phys. Rev. Letters*, **18**, 188 (1967).
17. H. S. Mani, Y. Tomazawa, and Y-P. Yao, *Phys. Rev. Letters*, **18**, 1084 (1967).

The Relations between PCAC, Axial-Charge Commutation Relations, and Conspiracy Theory*

STANLEY MANDELSTAM

*Department of Physics, University of California,
Berkeley, California*

1. Introduction

The object of this paper is to study some of the assumptions made in the theory of partially conserved axial currents and current commutators, and to investigate whether they may be derived from analyticity-unitarity assumptions or from one another. To begin with, we investigate the restrictions which PCAC and current-commutation relations impose on hadron scattering amplitudes. One such restriction is the Adler self consistency condition, which states that scattering amplitudes involving soft pions of low mass must be small. A further restriction is necessary for the self-consistency of the Adler-Weisberger relation. Expressed as a low-energy theorem, the Adler-Weisberger relation gives the axial-vector renormalization constant in terms of the antisymmetric part of the πN scattering amplitude. One can obtain similar relations between the axial-vector renormalization constant and the anti-symmetric part of the amplitude for the scattering of pions off any target at low energy. By eliminating the axial-vector renormalization constant, one can relate the antisymmetric part of the amplitude for scattering of low-energy pions by different targets. One easily finds that the antisymmetric part of the amplitude must be equal to a universal constant multiplied by the isotopic spin of the target. We shall refer to this relation as the Adler-Weisberger self-consistency condition.

All experimentally verifiable results of PCAC make use of the low mass of the pion, and it is uncertain whether the hypothesis of a partially conserved current has any content except in this approximation. One can only obtain exact results in a system where the pion mass is equal to zero, partial conservation then becomes exact conservation. We shall take the pion mass to be equal to zero throughout this paper, the results should be

* Research supported in part by the Air Force Office of Scientific Research, Office of Aerospace Research, under Grant No. AF-AFOSR-232-66.

true to a good approximation in nature, where the square of the pion mass is a good deal smaller than the square of the mass of any other particle.

Chew has made the suggestion that the Adler self-consistency condition might be a consequence of conspiracy theory,[1-3] which shows that several trajectories with different quantum numbers pass through the point $t = 0$ at equal values of α, or at values of α which differ by integers. From our point of view the important aspect of the theory is that it places restrictions on the Regge residues involving such trajectories. If the trajectory passes through an integral value of α at $t = 0$, as the pion trajectory does in our approximation, the Regge residues are products of two vertex constants associated with the pion. The Regge residue in a multiparticle reaction will be the product of two scattering amplitudes involving the pion. Conspiracy theory would therefore be expected to put a restriction on such amplitudes.

We shall show in Section 2 that the restrictions imposed by conspiracy theory do imply the Adler self-consistency condition, provided we assume that the conspiracy quantum number $|M|$ of the pion trajectory is one or greater. In outline, the argument will be that we are able to obtain a relation between the sense and nonsense residues associated with the pion trajectory. This relation shows that the sense amplitude is zero if $|M|$ is greater than the spin of the pion, i.e., if $|M|$ is greater than zero. Conspiracy theory only applies when *all four components of the pion momentum are zero*, and scattering amplitudes involving a soft massless pion therefore vanish. This is the Adler self-consistency condition. We shall mention briefly the arguments for believing that $|M|$ for the pion is equal to 1. Such arguments have been quoted by several authors independently of the relation between the value of $|M|$ and the Adler self-consistency condition.

Our object in Section 3 is to derive the Adler-Weisberger self-consistency condition from the Adler self-consistency condition. The antisymmetric part of the scattering amplitude is linear in the pion momenta if these momenta are small. Furthermore, by the Adler self-consistency condition, it has to vanish when either pion momentum is zero. These two requirements place strong restrictions on the amplitude. If we write the antisymmetric part of the amplitude in the form $T_\mu(p_1 - p_2)_\mu$, where the pion momenta p_1 and p_2 are small, we shall show that the restrictions imply that T_μ satisfies a divergence condition. The Adler-Weisberger self-consistency condition can then be proved following the arguments by which one proves that an electromagnetic field interacts with a conserved quantity.

The result of Sections 2 and 3 is therefore that all restrictions which PCAC or current commutators place on hadron scattering amplitudes can be obtained without mentioning currents if the pion has $|M| = 1$. We have not examined production amplitudes involving several soft pions, but it would be surprising if the situation were different for that case.

The next question, which will be discussed in Section 4, is whether the existence of massless particles satisfying the Adler self-consistency condition implies the existence of a conserved axial current. Our aim is to construct matrix elements of the axial current by solving Omnes-type equations. We are not concerned here with the deeper questions of solubility of the equations; we shall assume that Omnes equations such as those for vector-current matrix elements are soluble. There are, however, additional problems posed by axial-current conservation, since certain axial-current matrix elements are found to acquire a pole at $t = 0$ as a consequence of the conservation equation. In a theory with massless pions which satisfy the Adler self-consistency equation the poles can be attributed to one-pion intermediate states, and we shall find that everything is consistent. In other words, the existence of a conserved axial current in such a theory is on the same footing as the existence of a conserved vector current in a theory with the appropriate symmetry.

Finally we wish to treat the axial-current commutation relations, which we shall do in Section 5. Most successful applications of the commutation relations depend only on the commutator between total axial charges and not on the more detailed commutation relation between current densities. In particular, the Adler-Weisberger relation depends only on total-charge commutation relations. We shall show that the total-charge commutation relations can be derived from the conservation of the axial current without further assumption. One way to obtain the result would be to use the methods followed by Adler or Weisberger. They assumed knowledge of the commutation relations and related these to the antisymmetric part of the πN scattering amplitude, but it would be equally possible to assume the results of Section 3 and to deduce the commutation relations therefrom. This method assumes that the time-ordered product to two axial currents can be defined, and that it satisfies locality properties from which the reduction formulas can be obtained.

While we have no reason to doubt these assumptions we shall show that they are not necessary to derive the commutation relations. One can derive them directly from the properties of amplitudes involving soft pions, by examining the intermediate states involved in the commutator. We shall show that the only intermediate state which gives a contribution is that obtained from the initial or final state by the addition of one soft pion and we shall be able to calculate the contribution from this state explicitly. Our derivation will thus place the axial-charge commutation relations on a similar footing to the vector-charge commutation relations, they can be obtained from the conservation law by considering possible intermediate states. The type of intermediate state which contributes is different in the two cases. For the vector charge the intermediate states are the same as the initial states except possibly for an isotopic-spin rotation while, for the axial-charge, they differ from the initial states by the presence of a soft pion. The commutation relations are therefore closely bound up with the properties of scattering amplitudes involving soft pions and, in particular, with the Adler-Weisberger self-consistency condition.

Our axial charge commutation relations will of course involve an arbitrary constant, since the normalization of the charge is arbitrary. The normalization is usually *defined* so that this constant is unity. If we make the Gell-Mann universality assumption that such a normalization is appropriate for the weak interactions, we obtain the Adler-Weisberger relation in the usual way.

2. The Adler Self-Consistency Condition

In this section we wish to show that scattering amplitudes involving a zero-mass pion with conspiracy quantum number $|M|$ unequal to zero must necessarily satisfy the Adler self-consistency condition. Before doing so we make one or two comments on the reasons for believing that $|M|$ for the pion is in fact equal to 1.

We assume that the residue associated with the pion trajectory in nucleon-nucleon scattering does not vanish at $t = 0$, since measurements of backward proton-neutron scattering and of photoproduction of pions seem to require a nonzero residue. It then follows that the pion trajectory must be a member of one of the three types of conspiracy described by Freedman and Wang.[3] Their type I conspiracy has no pion trajectory, their type-II conspiracy a pion trajectory which conspires with an axial-vector trajectory, and their type-III conspiracy a pion trajectory which conspires with a scalar trajectory. If the mass of the pion is small, the axial trajectory of the type-II conspiracy would pass through $\alpha = 1$ at approximately the mass of the pion and would choose sense at this point. We would therefore have an axial-vector meson of mass approximately equal to that of the pion. Since no such particle is observed in nature, we reject a type-II conspiracy. With a type-III conspiracy, the scalar trajectory would pass through $\alpha = 0$ at approximately the mass of the pion, but it would choose nonsense at this point. It would therefore not give rise to any unobserved particle. Accordingly we assume that the pion trajectory is a member of a type-III conspiracy. Since a type-III conspiracy has $|M| = 1$, we conclude that this value of M is associated with the pion trajectory.[4]

We now return to the problem of obtaining the Adler self-consistency condition from the assumption that the pion is a member of a conspiracy with $|M| = 1$. Our starting-point is the relation between Lorentz poles and Regge poles. A Lorentz pole at position λ gives rise to a series of Regge poles at $\lambda - n$, where $n = 1, 2, 3 \ldots$.[5] The residue at the Regge pole factorizes in the form

$$\beta^n_{m_1 m_2, m_3 m_4} = \beta^n_{m_1 m_2} \beta^n_{m_3 m_4} \tag{2.1}$$

where m_1, m_2, m_3 and m_3 are the crossed-channel helicities of the particles. The factor $\beta_{m_1 m_2}$ is then given by the equation

$$\beta^n_{m_1 m_2} = \sum_s C(s_1, s_2, s, m_1, m_2, m) \gamma_s V^{M, \lambda, n}_{s, m} \tag{2.2}$$

where s_1 and s_2 are the spins of the two particles, m the total crossed-channel helicity ($m = m_1 - m_2$), and γ_s a constant related to the residue at the Lorentz pole. The factor V is a purely kinematical coefficient which depends on the properties of the groups $0(3, 1)$ and $0(2, 1)$ and which has been calculated by Sciarrino and Toller.[6] This coefficient is zero unless

$$|M| \leq |s| \tag{2.3}$$

We now observe that the right-hand side of (2) involves the helicity only in the known kinematical coefficient V. This feature is crucial to our result. It is valid only at $t = 0$, where the Lorentz-pole theory applies, at other values of t the dependence of the Regge residues on the helicity is not given by kinematical considerations. If λ is integral, so that the Regge trajectories pass through integral values of α at $t = 0$, the ratio between the sense and nonsense helicities will be determined by the kinematics. From this fact it will be possible to derive the Adler self-consistency condition.

We therefore have to investigate the dependence of the coefficient $V^{M, \lambda, n}_{s, m}$ on m. The pion trajectory passes through $\alpha = 0$ at $t = 0$. Since $\alpha = \lambda - n$ and we are interested in the leading trajectory, we conclude that $\lambda = 1$, $n = 1$. We have agreed to take $|M| = 1$. Furthermore, the Clebsch-Gordan coefficient in (2.2) vanishes unless $m \leq s$. We then find from Ref. 6, Eqs. (6) and (4.25), that

$$V^{\pm 1, 1, 1}_{s, m} = 0 \qquad m = 0 \tag{2.4a}$$

$$V^{\pm 1, 1, 1}_{s, m} \neq 0 \qquad |m| > 0 \tag{2.4b}$$

At $\alpha = 0$, the value $m = 0$ is the sense while all other values of m are nonsense. It follows that V is nonzero if and only if m has a nonsense value. This last result is valid for other values of M, λ and n, provided that the inequalities $|m| \leq s$, $|M| \leq s$ are satisfied, as they are for Lorentz poles, and provided also that $|M| \geq \lambda - n$.

From (2.2) and (2.4) we now conclude that either $\beta_{m_1 m_2}(t)$ is infinite for some $m > 0$ (if γ_3 is infinite) or $\beta_{m_1 m_2}(t) = 0$ for $m = 0$ (if γ_s is finite). It is not difficult to see that the first possibility must be rejected. The factor $\beta_{m_1 m_2}(t)$ could be infinite at $t = 0$ if there were fixed poles in the J or s planes, if there were singularities other than poles, or if two Regge trajectories intersected at $t = 0$. Fixed poles in the J-plane are excluded by unitarity, fixed poles in the s-plane by assumptions of analyticity in s and t, and singularities other than poles by our assumption of Regge asymptotic behavior. (The cuts in the angular momentum plane which are known to be present cannot result in β being infinite, as the discontinuities associated with them vanish when the angular momentum takes on an integral value of the correct signature.) The possibility of two trajectories intersecting cannot be rejected on *a priori* grounds, and we shall assume that this possibility is not realized. Even if it were, it is probable that our conclusions could still be obtained, but the argument would be rather more involved.

One might ask whether the coefficients γ_s on the right of (2.2) could be infinite, with the β_m's remaining finite for $|m| \geq 1$ as a result of a cancellation. By proceeding from large to small m and using the inequality $|m| \leq s$, one can easily reject this possibility. The Clebsch-Gordan coefficients vanish for $|m| \geq 1$ if $s = 0$, but this value of s is excluded by (2.3).

We thus conclude that $\beta(t)$ is zero when m takes on its sense value of zero. This means that the coupling constant associated with the vertex $AB\pi$ vanishes provided $m_A = m_B$. The equality of the masses is certainly a necessary condition for our reasoning to apply, since conspiracy theory, and in particular Eq. (2.2), only applies to the equal-mass zero-momentum-transfer amplitude, where all components of the pion momentum vanish individually.

We obtain the Adler self-consistency condition by extending this result to the case where A and B may be multiparticle systems. The reasoning of the last few paragraphs remains valid in this case, provided always that $m_A = m_B$. If for instance we consider the case where A consists of a nucleon and a second pion, B of a nucleon alone, the vertex $AB\pi$ becomes the πN scattering amplitude. We then obtain the result that the amplitude for the process $N_2 \pi_2 \to N_1 \pi_1$ vanishes if $m_{N_2 \pi_2} = m_{N_1}$ or, in other words, if all four components of the momentum of π_1 are zero. The point where this occurs is at the threshold for the physical region (provided $m_{\pi_2} = m_{\pi_1} = 0$). We emphasize again that our reasoning does not imply the vanishing of the amplitude when $m_{N_2 \pi_2} \neq m_N$, as conspiracy theory does not apply to the unequal-mass case. Our result is thus that the amplitude vanishes if the pion π_1 is soft, i.e., if all the components of its momentum are zero. This is precisely the Adler self-consistency condition.[7]

The assumption that $|M| \geq 1$ is crucial to our result. If M were equal to zero inequalities such as (2.4) would not be true.

In certain cases the results we have quoted are well known and follow from simple kinematics. The Regge residue factor for the process $A \to A$ + pion trajectory has a square-root zero at $t=0$ if $m_\pi = 0$. One can show this by making use of the usual kinematic-singularity analysis of helicity amplitudes[8] and, indeed, one can obtain a similar result for the process $A \to B\pi$, where A and B may be simple or composite systems with equal mass and the same intrinsic parity. The vanishing of the amplitude when A and B have opposite parities does not follow from simple kinematics and requires the assumption that the pion is a member of a conspiracy with $|M| = 1$. The Adler self-consistency condition applies to such cases, and, in particular, to the case where A is composed of B together with a soft pion. The Regge residue factor β for the process $A \to B$ + pion trajectory then has to behave like t at $t = 0$. A square-root zero would be excluded by the analyticity requirements.

3. The Adler-Weisberger Self-Consistency Condition

We begin this section by showing how one can eliminate the weak-interaction constants from the Adler-Weisberger equations for different processes and thereby obtain relations involving hadron amplitudes alone. We shall then proceed with our main purpose of showing that such relations can be derived from strong-interaction considerations alone, without referring to weak interactions or current commutators.

It is most convenient for our purposes to use the Adler-Weisberger equation in the form of a low-energy theorem. The amplitude for forward scattering of pions by nucleons at rest is written in the usual notation

$$A_{\alpha\beta} = \delta_{\alpha\beta} A^{(+)}(\nu) + \tfrac{1}{2}\varepsilon_{\alpha\beta\gamma}\tau_\gamma A^{(-)}(\nu) \tag{3.1}$$

where $A_{\alpha\beta}$ is a matrix in the charge space of the nucleon, ν the laboratory energy and τ the isotopic spin matrix. The function $A^{(-)}$ is usually referred to as the antisymmetric part of A, and the crossing relation implies that it is an odd function of ν. We may therefore write:

$$A^{(-)}(\nu) = a\nu + 0(\nu^2) \tag{3.2}$$

The Adler-Weisberger relation then states that

$$g^2 \frac{g_V^2}{g_A^2} = 2m^2 a \tag{3.3}$$

where g^2 is the πN coupling constant and $g_V{}^2/g_A{}^2$ the inverse of the axial-vector renormalization constant. Again we are taking the pion mass to be zero.

For the forward scattering of pions by any other particle at rest, one can write a slight generalization of (3.1):

$$A_{\alpha\beta} = \delta_{\alpha\beta} A^{(+,1)}(\nu) + \tfrac{1}{2}\{\rho_\alpha, \rho_\beta\} A^{(+,2)}(\nu) + \tfrac{1}{2}\varepsilon_{\alpha\beta\gamma} \rho_\gamma A^{(-)}(\nu) \qquad (3.4)$$

where the ρ's are the isotopic-spin matrices appropriate to the target particle. One can again expand $A^{(-)}$ in the form (3.2), and one can again derive (3.3). It therefore follows that the constant a must be the same as for πN scattering.

We therefore require that following consistency condition, which we shall call the Adler-Weisberger self-consistency condition: *If the amplitude for the forward scattering of a pion by a target particle is written in the form (3.4), and the antisymmetric part* $A^{(-)}$ *is expanded in powers of ν around $\nu = 0$, the coefficient of ν is equal to a universal constant* a *which is independent of the target particle.* We wish to derive this condition without the use of current commutators, we shall show that it is a consequence of the Adler self-consistency condition.

We take q_1 and $-q_2$ to be the momenta of the incoming and outgoing pion, p_1 and $-p_2$ to be the momenta of the incoming and outgoing target particle. Due to conservation of momentum there will be three independent four-momenta, which we take to be

$$Q = q_1 + q_2 = p_\pi + p_2 \qquad (3.5a)$$

$$q = \tfrac{1}{2}(q_1 - q_2) \qquad (3.5b)$$

$$p = \tfrac{1}{2}(p_1 - p_2) \qquad (3.5c)$$

For the forward scattering of massless particles at $\nu = 0$, both pions have zero four-momentum. We therefore expand the amplitude in powers of q:

$$T = T^{(0)}(p, Q) + q_\mu T_\mu{}^{(1)}(p, Q) + 0(q^2) \qquad (3.6)$$

We use the symbol T rather than A simply to indicate that we are expressing the amplitude as a function of the components of the momentum. We shall suppress the isotopic-spin indices α and β. By going to the laboratory system it is easily seen that

$$T_0{}^{(1)}(p_0, 0) = \tfrac{1}{2}\varepsilon_{\alpha\beta\gamma} \rho_\gamma a \qquad (3.7a)$$

where

$$p_0 = (m, 0, 0, 0) \qquad (3.7b)$$

and a is the constant in (3.2).

The Adler self-consistency condition can now be used to obtain a restriction on the amplitude $T^{(1)}$. We require that T vanish when either q_1 or q_2 is zero. It is of course easy to obtain an amplitude involving two powers of q which satisfies this restriction, but we wish to obtain an amplitude linear in q. Thus, putting $T = 0$ when $q_1 = 0$, i.e., when $q = -\frac{1}{2}Q$, we obtain

$$T^{(0)}(p, Q) - \tfrac{1}{2}Q_\mu T_\mu^{(1)}(p, Q) = 0(Q^2) \tag{3.8a}$$

Similarly, putting $T = 0$ when $q_2 = 0$, i.e., when $q = \frac{1}{2}Q$, we obtain

$$T^{(0)}(p, Q) + \tfrac{1}{2}Q_\mu T_\mu^{(1)}(p, Q) = 0(Q^2) \tag{3.8b}$$

Subtracting (3.8a) from (3.8b), we obtain the equation

$$Q_\mu T_\mu^{(1)}(p, Q) = 0(Q^2) \tag{3.9}$$

Thus $T_\mu^{(1)}$ *is a vector amplitude which satisfies the divergence condition when* Q *is small.*

We can treat the low-energy forward scattering of massless pions by a multiparticle system in a similar way.[9] We then take as our variables Q and q, defined by (3.5a) and (3.5b), together with a sufficient number of other momenta which we denote collectively by p. The development from (3.7) to (3.9) will remain valid, and again we find that the amplitude $T_\mu^{(1)}$ satisfies the divergence condition when Q is small.

The divergence condition essentially provides us with the result we require, since it is known that a vector interaction which satisfies the divergence condition when $Q = 0$ must involve a coupling constant which is proportional to a conserved quantity. This result has been derived from on-shell analyticity properties by Zwanziger[10] and by Weinberg.[11] They were interested in proving that the electromagnetic interaction is characterized by a conserved charge, but their methods are equally applicable to the present problem. Following their reasoning line by line, we can show that Eq. (3.9) requires a consistency condition which relates the amplitude for scattering of pions by different target particles. The condition is

$$T_{r,0}^{(1)}(p_0, 0) = c\sigma_r \tag{3.10}$$

where p_0 is defined by (3.7b), r is a subscript characterizing the target particle, σ_r is the matrix element, between the initial and final states of the target particle, of some conserved quantity, and c is a universal constant.

From (3.7a) we observe that T must transform under isotopic-spin rotations like the matrix $\varepsilon_{\alpha\beta\gamma}\rho_\gamma$. The only conserved quantity which does so is the matrix $\varepsilon_{\alpha\beta\gamma}\rho_\gamma$ itself. We must therefore identify σ_r with this matrix and, on comparing (3.10) with (3.2), we conclude that a is a universal constant independent of the target particle. This is the result we wished to prove.

4. Definition of a Conserved Axial Current

We now assume that we have a system containing a massless pseudo-scalar particle and we investigate whether it is possible to define a conserved axial current. It is not our aim to investigate the general problem of the solution of the dispersion-theoretic equations which define currents or form factors. We shall assume that such equations normally have solutions and shall concern ourselves with the particular problems raised by axial-current conservation. It will be found that such problems do not arise if the system contains a massless pseudoscalar particle and if the scattering amplitudes satisfy the Adler self-consistency condition.

As is well-known, a conserved axial current cannot exist in a theory which possesses neither chiral symmetry nor a massless pseudoscalar particle.[12] Indeed, it was the difficulties in this connection which led to the concept of Goldstone bosons. To indicate the nature of the problems involved, we begin by quoting the usual expression for the axial-vector form factor of the $N\overline{N}$ system:

$$j_\mu{}^5(q^2) = i\gamma_5\,\gamma_\mu\,f_1(q^2) + i\gamma_5\,\sigma_{\mu\nu}\,q_\nu\,f_2(q^2) + i\gamma_5\,q_\mu f_3(q^2) \qquad (4.1)$$

By using the Dirac equation for the nucleons, we easily find the following formula for the divergence of j:

$$q_\mu j_\mu{}^5(q^2) = 2m\gamma_5\,f_1(q^2) + i\gamma_5\,q^2 f_3(q^2) \qquad (4.2)$$

For conservation we therefore require

$$f_3(q^2) = (2mi/q^2)f_1(q^2) \qquad (4.3)$$

so that the formula for a conserved axial current becomes

$$j_\mu{}^5(q^2) = \gamma_5\{i\gamma_\mu - (2mq_\mu/q^2)\}f_1(q^2) + i\gamma_5\,\sigma_{\mu\nu}\,q_\nu\,f_2(q^2) \qquad (4.4)$$

The first term has a pole at $q^2 = 0$. We could eliminate this pole by demanding that $f_1(0) = 0$, but the matrix element of $j_\mu{}^5$ would then vanish when taken between nucleon states at rest. When discussing a conserved axial current one usually implies that this matrix element does not vanish, and we shall assume that $f_1(0) \neq 0$.

On taking the matrix element of (4.4) between states at rest, we find:

$$\langle j_i{}^5(0)\rangle = \{\sigma_i - [(\boldsymbol{\sigma q})q_i/\boldsymbol{q}^2]\}f_1(0) \qquad (4.5a)$$

$$\langle j_0{}^5(0)\rangle = 0 \qquad (4.5b)$$

the pole in (4.4) therefore causes no infinity in the matrix element, but it does cause a discontinuity at $q = 0$. In order for this discontinuity to disappear $f_1(0)$ would have to be equal to zero.

One can generalize this result to apply to the axial-vector form factor of a particle with arbitrary spin by making use of the analyticity properties of helicity amplitudes. It follows from the results of Ref. 8 that the axial-vector form factor of any particle must vanish like q^2 as q^2 approaches zero. If this is not the case the invariant form factors must have a dynamical pole at $q^2 = 0$, analogous to the pole in (4.4).

If a theory allows the definition of a conserved axial current, it must therefore have massless particles in order to produce the dynamical pole. From (4.4) we observe that the coupling between the axial current and the massless particle must have the form

$$bq_\mu \qquad (4.6)$$

where b is a constant, so that the particle must be pseudoscalar. We shall henceforth refer to it as a pion. If the pion is coupled to a particle such as the nucleon by a matrix of the form $iG\Gamma_5$, the axial-vector form vector will contain a term

$$-bG(q_\mu/q^2)\Gamma_5 \qquad (4.7)$$

The Feynman diagram corresponding to such a term is shown in Fig. 1.

It is not difficult to see that scattering amplitudes involving the pion must satisfy the Adler self-consistency condition. The invariants associated with the axial-vector form factor between two systems of *equal mass but different parity* will have no pole at $q^2 = 0$. Such invariants have exactly the same properties as those for the vector form factor between two systems of the same parity. On the other hand, equation (4.7) indicates that a pole is present unless $G = 0$. We therefore conclude that the coupling of the pion between two systems of equal mass but opposite parity vanishes, this is the Adler self-consistency condition.

If our system contains massless particles whose scattering amplitudes satisfy the Adler self-consistency condition, so that poles in the form-factor

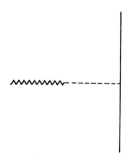

FIG. 1. Pole diagram in a matrix element for a current.

invariants are allowed, the problem of finding a conserved axial-vector current is very similar to the analogous problem involving the vector current. In general, an axial-vector form factor such as that given in (4.1) will consist of two parts, one being divergenceless and the other the gradient of a pseudoscalar. We shall refer to these parts as the conserved part and the gradient part. By writing dispersion relations in q^2 and using unitarity, one obtains Omnes-type equations for the form factors. Furthermore, the unitarity equations will not mix the conserved part with the gradient part, as the first involves intermediate states with $J = 1$, the second intermediate states with $J = 0$. One therefore obtains Omnes-type equations for functions such as f_1 and f_2 in (4.4), and they can be treated in the same way as the corresponding equations for the vector form factors.

The value of a form factor at $q = 0$ will be fixed once the constant b of (4.6) and the appropriate coupling constant G are known. The residue of the pole in the invariant function will be given by (4.7) and, if this residue is known, one can find the form factor at $q = 0$ by an equation such as (4.4). Again the situation is similar to that we encounter with a conserved vector current, when the form factor at $q = 0$ is determined by the conservation laws. The situations are not identical since, with a vector current, the form factor at $q = 0$ depends only on the quantum numbers of the relevant particle, whereas with an axial current the form factor depends on coupling constants involving the pion.

The normalization of the current is not defined by the Omnes equations, which are linear. The constant b in (4.6) will remain undetermined until the current is normalized. One usually normalizes the axial current through the weak-interaction effective Lagrangian, in which case the function f_1 in (4.4) will be equal to g_A/g_V at $q^2 = 0$. Comparing the pole terms in (4.4) and (4.7), and putting $G = g$, $\Gamma_5 = \gamma_5$ in the latter equation, we then find that

$$b = 2mg_A/gg_V \qquad (4.9)$$

The matrix element of the axial current between the one-pion state and the vacuum will be equal to

$$(-i/2\pi^{3/2})(1/2p_0)^{1/2}bp_\mu \qquad (4.10)$$

where p is the momentum of the pion. If we assume that this formula is approximately true when the pion mass is small but not zero, and if we use Eq. (4.9) for b, we can obtain the Goldberger-Treiman relation in the usual way.

5. Commutation Relations between Axial Charges

In a theory with a conserved vector current the commutation relations between total charges follow from the conservation equations, it is unnecessary to make a separate assumption. We shall show in this section that the same is true of the commutation relations between total axial charges in a theory with a conserved axial current. The intermediate states involved in the axial-charge commutator will be different from those involved in the vector-charge commutator, since the massless pions play an essential role in the conservation of the axial current, but the final result is similar.

As we explained in Section 1, we wish to obtain our result by showing that only a small number of states can contribute to the commutator, and that the matrix elements involving such states can be written down explicitly.

The equal-time commutator between two charge densities is given by the matrix element

$$(2\pi)^2 \int dq_{10}\, dq_{20} \langle N \,|[j_{0\alpha}^{\ 5}(\mathbf{q}_1, q_{10}), j_{0\beta}^{\ 5}(\mathbf{q}_2, q_{20})]|\, N \rangle$$

$$\mathbf{q}_1 = \mathbf{q}_2 = 0 \tag{5.1}$$

where N indicates a nucleon state (or any other state) at rest, and $j_{0\alpha}^{\ 2}$ is the zeroth component of the Fourier transform of the charge density.[13] The subscripts α and β refer to the isotopic spin. With $\mathbf{q}_1 = \mathbf{q}_2 = 0$, axial current conservation would imply that $j_{0\alpha}^{\ 5}$ had a factor $\delta(q_0)$, and the integrations over q_{10} and q_{20} in (5.1) would be trivial. We shall, however, find it necessary to use a limiting procedure in which $\mathbf{q}_1$ and $\mathbf{q}_2$ tend to zero. We therefore leave (5.1) as it stands.

One can now insert a complete set of intermediate states in (5.1). If $\mathbf{q}_1 = \mathbf{q}_2 = 0$, the divergence condition $q_\mu j_\mu(q) = 0$ eliminates the intermediate states with energy different from that of the nucleon. On the other hand, the one-nucleon intermediate state itself will not contribute, since we saw in the previous section that the expectation value of $j_0^{\ 5}(0)$ for the one-nucleon state is zero. The only other intermediate states which can possibly contribute is a state consisting of a nucleon and a soft pion. We shall show that the matrix element involving such an intermediate state is singular, and that the state gives a finite contribution to the commutator.

Two singular diagrams for the matrix element of the axial charge. between a nucleon state and a nucleon-soft-pion state are shown in Fig. 2.

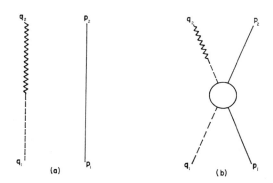

FIG. 2. Important diagrams for a matrix element of an axial charge.

In Fig. 2a, the pion changes into a current while the nucleon goes on un-changed, in Fig. 2b we encounter a pion pole term of the type discussed in the previous section. Owing to the singular nature of the quantities involved we shall consider the momenta in (5.1) to be small but nonzero, when we have found the commutator we shall allow the momenta to approach zero. The matrix elements associated with the diagrams are:

$$\langle N(\mathbf{p}_2)|j_{\mu,\alpha}^5(\mathbf{q}_2\,q_{20})|\,N(\mathbf{p}_1),\,\pi_\beta(\mathbf{q}_1)\rangle_1$$

$$= (2\pi)^{1/2}b\,\frac{q_{2\mu}}{(2q_{20})^{1/2}}\,\delta^3(\mathbf{q}_2-\mathbf{q}_1)\delta(q_{20}-|\mathbf{q}_1|)\delta^3(\mathbf{p}_2-\mathbf{p}_1)\delta_{\alpha\beta} \qquad (5.2\text{a})$$

$$\langle N(\mathbf{p}_2)|j_{\mu,\alpha}^5(\mathbf{q}_2\,q_{20})|\,N(\mathbf{p}_1),\,\pi_\beta(\mathbf{q}_1)\rangle_2$$

$$= -\,\frac{b}{(2\pi)^{5/2}}\,\frac{1}{(2q_{10})^{1/2}}\,\frac{q_{2\mu}}{q_{20}^{\,2}-\mathbf{q}_2^{\,2}}\,\delta^3(\mathbf{p}_2+\mathbf{q}_2-\mathbf{p}_1-\mathbf{q}_1)$$

$$\times\,\delta(q_{20}-|\mathbf{q}_1|)A_{\alpha\beta}(|\mathbf{q}_1|)$$

$$\mathbf{p}_1,\,\mathbf{p}_2,\,\mathbf{q}_1,\,\mathbf{q}_2,\,q_{20}\ \text{small} \qquad (5.2\text{b})$$

The subscripts 1 and 2 on the matrix elements on the left of (5.2) indicate that they correspond to Figs 2a and 2b, respectively. In writing down (5.2a), we have assumed that $p_{10} = p_{20} = M$, apart from terms of second order which we have dropped.

We draw attention to the factor $1/(q_{20}^{\,2}-\mathbf{q}^2)$ in (5.2b), which becomes large when q_{20} and $\mathbf{q}$ are small. There will of course be other diagrams for the matrix elements in (5.2), but they will be negligible compared with (5.2b) when $\mathbf{q}_{20}$ and q are small.

We can now evaluate the matrix element $\langle N|j_\alpha^5 j_\beta^5|N\rangle$. With πN intermediate states, the matrix element becomes $\langle N|j_\alpha^5|N\pi\rangle\langle N\pi|j_\beta^5|N\rangle$.

Each of the two factors will be the sum of two terms corresponding to (5.2a) and (5.2b). The term where both factors correspond to (5.2a) will only be nonzero when $\alpha = \beta$. Since we are interested in finding the commutator of j_α and j_β, we shall restrict ourselves to the case $\alpha \neq \beta$. In the term where both factors correspond to (5.2b), there is an integration over q_{10} and q_{20} (see Eq. (5.1)), this integration removes the singular factor $1/(q_{20}^2 - \mathbf{q}^2)$. The amplitude $A(|\mathbf{q}_1|)$ in (5.2) involves soft pions and is therefore small and, as may easily be verified, the result is that the whole term is small.

We are left with terms where no factor corresponds to (5.2a) and one to (5.2b). The integrations over $\mathbf{q}_{01}$ and q_{02} in (5.1), as well as the integration over the intermediate-state pion three-momenta, are trivial because of the δ-functions in (5.2a) and (5.2b). We thus obtain the equation

$$(2\pi)^2 \int dq_{10}\, dq_{20}\, d^3\mathbf{q}' \sum_\gamma \langle N(\mathbf{p}_2)|j_{0\alpha}{}^5(\mathbf{q}_2, q_{20})|N(\mathbf{p}_2 + \mathbf{q}_2 - \mathbf{q}'(\pi_\gamma(\mathbf{q}'))\rangle_1$$

$$\times \langle N(\mathbf{p}_2 + \mathbf{q}_2 - \mathbf{q}')\pi_\gamma(\mathbf{q}')|j_{0\beta}{}^5(\mathbf{q}_1, q_{10})|N(\mathbf{p}_1)\rangle_2$$

$$= \frac{1}{2} b^2 \frac{|\mathbf{q}_2|}{\mathbf{q}_2{}^2 - \mathbf{q}_1{}^2} A_{\alpha\beta}(|\mathbf{q}_2|)\delta^3(\mathbf{p}_1 + \mathbf{q}_1 - \mathbf{p}_2 - \mathbf{q}_2) \qquad (5.3a)$$

The corresponding term, with the subscripts 1 and 2 interchanged, is

$$\frac{1}{2} b^2 \frac{|\mathbf{q}_1|}{\mathbf{q}_1{}^2 - \mathbf{q}_2{}^2} A_{\alpha\beta}(|\mathbf{q}_1|)\delta^3(\mathbf{p}_1 + \mathbf{q}_1 - \mathbf{p}_2 - \mathbf{q}_2) \qquad (5.3b)$$

In evaluating the amplitude $A_{\alpha\beta}$ in (5.3a) and (5.3b) we shall ignore the symmetric part, as it will not contribute to the commutator of $j_{0\alpha}$ and $j_{0\beta}$. From (3.1) and (3.2) we may therefore write

$$A_{\alpha\beta}(|\mathbf{q}|) = \tfrac{1}{2} a \varepsilon_{\alpha\beta\gamma}\, \tau_\gamma\, |\mathbf{q}| \qquad (5.4)$$

Substituting (5.4) and (5.3) and adding (5.3a) and (5.3b), we are led to the result

$$(2\pi)^2 \int dq_{10}\, dq_{20} \langle N(\mathbf{p}_2)|j_{0\alpha}{}^5(\mathbf{q}_2, q_{20}) j_{0\beta}{}^5(\mathbf{q}_1, q_{10})|N(\mathbf{p}_1)\rangle$$

$$= \tfrac{1}{2} b^2 \tfrac{1}{2} a \varepsilon_{\alpha\beta\gamma}\, \tau_\gamma\, \delta^3(\mathbf{p}_1 + \mathbf{q}_1 - \mathbf{p}_2 - \mathbf{q}_2)$$

$$\mathbf{p}_1, \mathbf{p}_2, \mathbf{q}_1, \mathbf{q}_2 \text{ small} \qquad (5.5)$$

Equation (5.5), together with the corresponding equation in which the order of $j_{0\alpha}{}^5$ and $j_{0\beta}{}^5$ is reversed, give us the commutation relation. One can write similar equations for any initial and final states, the matrix τ_γ being replaced by the more general isotopic-spin matrix ρ_γ. Since the initial and final states are arbitrary, we would be tempted to write (5.5) as an operator equation. However, we have not considered the possibility that the initial and final states themselves contain soft pions, which may interact directly with the currents as in Fig. 3. Figure 3a corresponds to an intermediate state with two pions, Fig. 3b to an intermediate state with none. It is not difficult to see that the contribution of Fig. 3a to the matrix element $\langle j_\alpha^5 j_\beta^5 \rangle$ is equal to the contribution of Fig. 3b to the matrix element $\langle j_\beta^5 j_\alpha^5 \rangle$. Such diagrams therefore contribute to the product of two currents but not to their commutator.

One may also enquire about diagrams such as Fig. 4, in which only one pion intersects directly with a current. The right half of this diagram is the matrix element of an axial current between states at rest. As before, this matrix element is only nonzero when B consists of A together with a soft pion, i.e., a pion whose momenta are of the same order of magnitude as $\mathbf{p}_1$, $\mathbf{p}_2$, $\mathbf{q}_1$ and $\mathbf{q}_2$. The phase space associated with the states B which fulfil this condition is of the order of magnitude p^3 and, even if the right half of Fig. 4 contains a pole term such as Fig. 2b, it is fairly easy to see that the small phase space renders the process unimportant in the limit of vanishing $\mathbf{p}_1$, $\mathbf{p}_2$, $\mathbf{q}_1$ and $\mathbf{q}_2$. The amplitude associated with Fig. 4 will therefore not contribute to the axial-charge commutator.

The presence of soft pions in the initial and final states of (5.5) will thus give rise only to terms which cancel when we take the commutator, and we can write the commutator equation as an operator equation. The matrix τ_γ on the right of (5.5), or its generalization ρ_γ, is just the matrix element of the vector charge between the initial and final states. We may therefore write

$$(2\pi)^2 \int dq_{10}\, dq_{20} [j_{0\alpha}^5(\mathbf{q}_2, q_{20}) j_{0\beta}^5(\mathbf{q}_1, q_{10})] = b^2 \tfrac{1}{2} a \varepsilon_{\alpha\beta\gamma} j_{0\gamma}(\mathbf{q}_1, 0)\, \delta^3(\mathbf{q}_1 - \mathbf{q}_2)$$

$$(5.6)$$

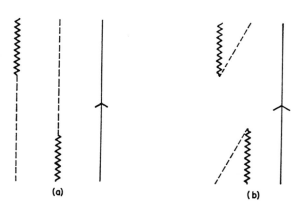

(a)   (b)

FIG. 3. Diagrams for the matrix element of an axial charge between states containing soft pions.

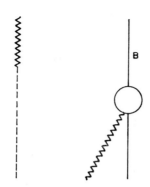

FIG. 4. A diagram which does not contribute to the matrix element of an axial charge.

This is the usual equal-time commutation relation between total axial charges, apart from the factor $\frac{1}{2}ab^2$.

One cannot get rid of the numerical factor on the right of (5.6), since the normalization of the current is not defined. One can normalize the current so that the constant $\frac{1}{2}ab^2$ is equal to unity, i.e., so that

$$b^2 = 2/a \qquad (5.7)$$

Without further assumption one must regard (4.9) and (5.7) as alternative normalizations of the constant b. The Gell-Mann universality assumption is that the axial current, when normalized so that the factor $\frac{1}{2}ab^2$ in (5.6) is unity, enters into the weak-interaction Lagrangian with the same coupling constant as the vector current. With that assumption the normalizations (5.7) and (4.9) are the same and, by eliminating b from the two equations, we obtain the Adler-Weisberger relation

$$g^2 g_V^2/g_A^2 = 2m^2 a \qquad (5.8)$$

Nothing in the arguments of Section 3 or this section prevents the constant a from being zero. In that case the axial charges would commute, as they do in certain models. There is no theoretical reason for preferring the value $a = 0$ to any other value of a, and experimentally a is not equal to zero.

To summarize the contents of this section, the conservation equation shows that most intermediate states give no contribution to the matrix element of the commutator of two axial charges. The important intermediate states are those obtained from the initial state by the addition of one soft pion. In this respect the axial charge is different from the vector charge, where the important intermediate states differ from the initial states only by an isotopic-spin rotation. The difference is due to the fact that conservation of axial charge does not correspond to a symmetry of the system in the usual sense. Nevertheless, the commutation relations have their expected form. They are intimately connected with the Adler-Weisberger self-consistency condition on the anti-symmetric part of an amplitude for the forward scattering of a soft pion.

6. Concluding Remarks

The arguments of Sections 3 and 4 show that all restrictions which PCAC and current commutators have so far provided on hadron scattering amplitudes can also be obtained from analyticity and unitarity, always on the assumption that the pion trajectory has $|M| = 1$. This is not to say that the concepts of PCAC and current commutators play no useful role. Indeed, it was through them that attention was directed to the Adler and Adler-Weisberger self-consistency conditions in the first place. However, the assumption of PCAC and axial-charge commutation relations appears to be an alternative to part of the content of the usual strong-interaction assumptions rather than an addition to them.

In this connection it is interesting to note that Gilman and Harari,[14] in their attempt to obtain correlations between resonance parameters from superconvergence relations and current commutators, really only use the current commutators to obtain low-energy theorems of the type discussed in this paper. Their results may therefore be considered as consequences of the usual analyticity and unitarity assumptions.

If one wishes to obtain results related to the conserved axial current itself, as opposed to results related to hadron scattering amplitudes, one has of course to make certain assumptions beyond on-shell analyticity. The usual assumption is that a matrix element of a current has the appropriate analyticity and unitarity properties. We are all familiar with such assumptions applied to the vector current and, if one applies them to the axial-vector current, one can put PCAC on the same footing as ordinary current conservation. Furthermore, one can derive the commutation relations for total charge from the conservation of the axial current.

As in the case of the vector charge, the method of obtaining the axial-charge commutation relations is to examine the intermediate states involved. Only a small number of intermediate states give a nonzero contribution. The actual intermediate states which come into play with the axial-charge commutator are different from those which come into play with the vector-charge commutator, owing to the intimate connection of the axial-current conservation with soft pions. The details has been discussed in Section 4.

The vector and axial-vector charges satisfy the commutation relations of the $SU(2) \times SU(2)$ algebra. Nevertheless, $SU(2) \times SU(2)$ is not a symmetry of the system in the usual sense, the states of the system are multiplets of $SU(2)$ only. Our system is the analogue of canonical field-theoretic model in which the Lagrangian possesses a higher symmetry than the system itself. Such a system must have Goldstone bosons, and it has been known for some time that a massless pion plays the role of a Goldstone boson in a system with a conserved axial current. The existence of such bosons can be studied without reference to a Lagrangian.[1][2] If a current satisfies a conservation law but the system does not possess the symmetry appropriate to that conservation law, Goldstone bosons must exist. The current commutation relations can then correspond to a larger algebra than the symmetry algebra of the system. In nature it appears that the existence of conserved charges satisfying the $SU(2) \times SU(2)$ algebra is true to a fairly high degree of approximation, whereas $SU(2) \times SU(2)$ symmetry, if it has any meaning at all, is badly broken.

It is important to stress the different footing on which total-charge commutation relations and current-density commutation relations stand. The former are a consequence of conservation equations, which in turn are a consequence of symmetry principles (for the vector charges) or of the existence of a massless particle with $|M| \neq 0$ (for the axial charge). The current-density commutation relations represent a further assumption which cannot be obtained from such reasoning. Furthermore, in all applications of current-density commutation relations to experiments which are practicable in the forseeable future, one has to make assumptions regarding the validity of taking the limit $p \to \infty$ or, alternatively, assumptions regarding unsubtracted dispersion relations for weak amplitudes. Questions regarding the validity of such assumptions, or of the relationship of their validity to the bootstrap assumptions, have not been answered as yet. As we have shown in this paper, most of the sum rules for which we have experimental evidence are consequences of much more general assumptions and have no bearing on such questions. The one exception is the Cabibbo-Radicatti rum rule. Further evidence in its favor, or evidence regarding other current-density sum rules so far untested, will provide answers to questions which go beyond the general arguments of this paper.

I have benefitted greatly from discussions with K. Bardakci and G. F. Chew.

References

1. D. V. Volkov and V. N. Gribov, *Zh. Eksperim. i. Teor. Fiz.*, **44**, 1068 (1963) (English transl.; *Soviet Physics—JETP*, **17**, 720 (1965)).
2. M. Toller to be published.
3. D. Z. Freedman, and J. M. Wang, *Phys. Rev.*, **160**, 1560 (1967).

4. Another possibility is that the residue associated with the pion trajectory does vanish at $t = 0$, but that two other conspiring trajectories pass through $\alpha = 0$ or $\alpha = 1$ near $t = 0$. It has been found possible to fit the data with such trajectories, but they usually imply the existence of particles which have not been seen. I should like to thank Dr. G. Ringland for discussions on this point.

5. In our definition of λ we are following Toller's notation. The relation between λ and the n of Freedman and Wang is $\lambda = n + 1$.

6. A Sciarrino and M. Toller, *J. Math. Phys.*, **8**, 1252 (1967).

7. If we are interested in the behavior of multiparticle amplitudes involving a soft pion, we must apply special treatment to the bremsstrahlung diagrams. Owing to the one-particle pole, the nonsense amplitudes involving such diagrams are infinite and the sense amplitudes are finite. The bremsstrahlung diagrams will only affect P-wave pions and will involve the pion momentum in the form $\mathbf{q}/|\mathbf{q}|$. Our results will still apply to S-wave pions. I should like to thank S. Weinberg and J. Weis for discussions on this point.

8. L.-L. Wang, *Phys. Rev.*, **142**, 1187 (1966).

9. To avoid the bremsstrahlung diagrams we should work with the amplitude from which the P-wave pion states have seen projected out.

10. D. Zwanziger, *Phys. Rev.*, **133**, B1036 (1964). .

11. S. Weinberg, *Phys. Rev.*, **135**, B1049 (1964).

12. See, for instance, Y. Nambu and G. Jona-Lasinio, *Phys. Rev.*, **122**, 345 (1961).

13. We normalize the Fourier transforms in the usual way, i.e.,
$$j(p) = (2\pi)^{-2} \int d^2x e^{-ipx} J(x).$$
Feynman diagrams will then have a factor $(2\pi)^{-2}$ associated with an external wavy line.

14. F. Gilman and H. Harari, *Phys. Rev. Letters*, **18**, 1150 (1967); *ibid*, **19**, 723 (1967), and to be published.

Excited Nucleons
and the Baryonic Supermultiplets

(Introductory Talk in the Panel Discussion Session)

R. H. DALITZ

*Department of Theoretical Physics, Oxford University,
Oxford, England*

The particular interest of πN resonant states lies in their role as convenient indicators for the spin and parity of the baryonic $SU(3)$ multiplets. The πN resonant states are now readily accessible. Pion beams of suitably high intensity are available at many accelerator laboratories, and the availability of polarized proton targets means that rather accurate and complete data can now be obtained conveniently for the scattering and polarization angular distributions for $\pi^{\pm}p$ interactions (including the charge-exchange process) at a definite value for the c.m. energy. At the same time, the art of analyzing such data in terms of partial-wave amplitudes has been developing rapidly. In this volume, Lovelace and his collaborators have reported a new phase-shift analysis[1-4] up to about 2000 MeV c.m. energy, indicating a considerable number of new resonant states in this mass region, as listed in Table I. This work represents a major step forward in our knowledge of baryonic resonance states.

The present evidence on the excited baryon multiplets indicates that these are limited to unitary singlets (the $Y_0{}^*$ states), to unitary octets (B^*, consisting of N^*, Λ^*, Σ^*, Ξ^* states), and to unitary decuplets D^*, consisting (Δ^*, $Y_1{}^*$, Ξ^*, Ω^* states). Evidence for bumps in the K^+p and K^+d total cross-section curves, the so-called $Z_0(1865)$ and $Z_1(1900)$ bumps, was first reported by Cool et al.[5] and later confirmed by Bugg et al.[6] Abrams et al.[7] have recently reported more detailed observations which show that the $Z_1(1900)$ bump itself has structure, the total cross section having two more, much smaller, bumps on the high-energy side of $Z_1(1900)$. However, the evidence available on K^+p interactions near the main $Z_1(1900)$ bump indicates that it is not due to a resonant state but arises from characteristic properties of the $KN \rightarrow K\Delta$ inelastic cross section, due to coherent interference between the amplitudes in several partial waves. There is rather little evidence available on the scattering and reaction processes

TABLE I

The πN Resonance States Given by the Phase-Shift Analysis of
πN scattering and Polarization Data up to 2000 MeV c.m. Energy[1,3,†]

N^* states				Δ^* states			
Wave	Mass	Γ(MeV)	Γ_{el}/Γ	Wave	Mass	Γ(MeV)	Γ_{el}/Γ
P_{11}	939	—	—	P_{33}	1236	125	1.0
P_{11}	1470	210	0.65	S_{31}	1640	180	0.30
D_{13}	1520	115	0.55	P_{33}	1688	280	0.10
S_{11}	1535	120	0.35	D_{33}	1691	270	0.14
D_{15}	1680	170	0.40	F_{35}	1913	350	0.16
D_{13}	1675	?	?	P_{31}	1934	340	0.30
F_{15}	1690	130	0.65	F_{37}	1950	220	0.40
S_{11}	1710	300	0.80	D_{35}	1954	310	0.15
P_{11}	1751	330	0.32				
P_{13}	1860	300	0.21				
F_{17}	1983	225	0.13				
D_{13}	2057	290	0.26				
G_{17}	2200	300	0.35				

† The table also includes the state $D_{13}(1675)$, whose existence was suggested
by the structure found in Ref. 1 for $\eta(D_{13})$ for pion lab. kinetic energy 900 MeV
and which has been confirmed in more recent work.[4]

occurring near the $Z_0(1865)$ bump, and its interpretation (and the interpretation of the K^+d total cross section in terms of $I = 0$ KN interactions)
is confused by the uncertainties involved in making allowance for the
binding of the target neutron in the deuterium. Here we shall assume that
this $Z_0(1865)$ bump also represents an effect of some threshold phenomenon
complicated by the nuclear structure effects in deuterium.

With this situation, then, that excited baryons consist only of Y_0^*,
B^*, and D^* states, each N^* state observed corresponds to a baryonic
octet B^* and each Δ^* state observed corresponds to a baryonic decuplet
D^*. The spin-parity for each N^* or Δ^* state is determined directly by the
πN partial-wave analysis, and this determines at once the spin-parity for
the corresponding octet B^* or decuplet D^*. To date, only two singlet Y_0^*
states have been clearly established, $Y_0^*(1405)$ and $Y_0^*(1520)$. Such
Y_0^* states are studied most directly from observations on the K^-p interaction processes, or from observations on their decay processes (especially the
processes $Y_0^* \to \pi^\pm \Sigma^\mp$ and $\Lambda\eta$) following Y_0^* production in high-energy
collisions. The study of the decay properties for $Y_0^*(1520)$, following the

formation process $K^-p \to Y_0^*(1520)$ have led to the result $(\frac{3}{2}-)$ for its spin-parity; $Y_0^*(1405)$ has spin-parity $(\frac{1}{2}-)$, a result deduced indirectly from the properties of the s-wave K^-p scattering and reaction processes just above the $\overline{K}N$ threshold. We note here that the quark model for hadrons requires the (low-lying) baryonic states to result from the binding of three quarks, and that three-quark systems can form only unitary singlet, octet, and decuplet states. In this three-quark model, the low-lying excited baryon states [beyond the low-lying $(\frac{3}{2}+)$ decuplet] are generated by excitations of the internal orbital motions of the quarks.

Table I lists all the nucleonic resonances known up to mass value 2200 MeV. These states are characterized by the orbital angular momentum L in the πN channel, given by $S, P, D, \ldots$ for $L = 0, 1, 2, \ldots$, the isospin I (the first suffix being $2I$) and the total angular momentum J (the second suffix being $2J$). To each N^* state, $L_{1,(2J)}$, there corresponds a baryonic octet with spin-parity $(J, (-1)^{L+1})$; to each Δ^* state, $L_{3(2J)}$, there corresponds a baryonic decuplet with spin-parity $[J, (-1)^{L+1}]$. Some of the Y^* states which have been established (Λ^* and Σ^* for the octets, Y_1^* for the decuplets)[8] are known to correspond to particular N^* and Δ^* states among those listed in Table I (e.g., $\Lambda^*(1815)$ and $\Sigma^*(1910)$ correspond to the $F_{15} N^*(1690)$), but there are far fewer Y^* states known than are required to fill out all of these octets and decuplets. Only four Ξ states are known, the $(\frac{1}{2}+)$ baryon $\Xi(1320)$, the $(\frac{3}{2}+)$ decuplet state $\Xi^*(1530)$, and two others $\Xi^*(1815)$ and $\Xi^*(1930)$, whose spin-parity values have not yet been established. The only Ω state known is the $(\frac{3}{2}+)$ decuplet state $\Omega(1672)$.

The indications are that the *superstrong forces* which give rise to these baryonic states obey $SU(6)$ symmetry, in some fair approximation. With the three-quark model for these states, this suggests that the q–q potentials should obey $SU(6)$ symmetry in the static limit, where the internal motions of the quarks are nonrelativistic, at least for the low-lying baryonic states. This approximate higher symmetry means that $SU(3)$ multiplets with different spin values (but the same parity) may themselves be grouped into supermultiplets, related with the $SU(6)$ symmetry. The basic $SU(6)$ representations which can be appropriate for a three-quark system are limited to the following:

$SU(6)$ representation (dimensionality)	56	70	20
Permutation symmetry	S	M	A

The decompositions of these $SU(6)$ representations into unitary-spin multiplets and (Pauli) spin states [i.e., according to the base states of the product group $SU(3) \times SU(2)$] are given in Table II. For example, the baryon octet $(\frac{1}{2}+)$ and the baryonic decuplet $(\frac{3}{2}+)$ are believed to form

TABLE II

The Decomposition of the $SU(6)$ Representations into
$SU(3)$-Multiplets $\{\alpha\}$ with Spin S

$SU(6)$ representation	$(\{\alpha\}, 2S + 1)$ reduction
20	$(\{8\}, 2) + (\{1\}, 4)$
70	$(\{1\}, 2) + (\{10\}, 2) + (\{8\}, 2) + (\{8\}, 4)$
56	$(\{8\}, 2) + (\{10\}, 4)$

together the basis for a 56-dimensional representation of $SU(6)$ symmetry;
thus

$$(56, L = 0+) \rightarrow (\{8\}, {}^2S_{1/2}(+)) + (\{10\}, {}^2S_{3/2}(+)) \qquad (1)$$

Here, we have adopted the notation $(\{\alpha\}, {}^{2S+1}L_J(w))$ for the description
of an $SU(3)$ multiplet belonging to representation $\{\alpha\}$, derived from a three-
quark system with spin S and total orbital angular momentum L, coupled
to total spin J, the total parity being w.

In this paper, we wish to point out that all of the levels listed in Table I,
together with the two Y_0^* singlet states, can be assigned to $SU(6)$ supermultiplets with $(L, w) = 0+$, $1-$, or $2+$, in a natural way, with a small
number of states outstanding at the upper end of the mass range, which are
consistent with the $SU(6)$ supermultiplets to be expected to occur just
beyond 2000 MeV. Further, the sequence of orbital excitations $L = 0+$,
$1-$, $2+$ and of $SU(6)$ representations follows naturally the sequence
expected for the shell-model states appropriate to a three-quark system.
This model for the baryonic states was first proposed by Greenberg[9] in
1964, who tabulated all the $SU(3)$ multiplets which are contained within
the shell-model states corresponding to at most three-quantum excitation.

We now consider the series of harmonic states $(1s)$, $(1p)$, $(2s)$, $(1d)$, ...
We assume $SU(6)$ symmetry for the superstrong forces. The energy of a
given shell-model state will depend first on the degree of excitation of the
shell-model configuration, then on both the $SU(6)$-representation α and
the details of the internal orbital motions. For a given configuration, we
generally expect the state with greatest permutation symmetry to lie lowest
in energy, at least for simple q–q forces which are attractive for relative
s-wave motion. Owing to the existence of forces which do not obey $SU(6)$
symmetry, each $SU(6)$ supermultiplet will split into $SU(3)$ multiplets, with
masses depending on the $SU(3)$ representation label $\{\beta\}$, the internal spin S
and the total spin J. In turn, owing to the moderately strong $SU(3)$-
breaking forces, these $SU(3)$ multiplets are themselves split into isospin

multiplets with definite hypercharge Y and isospin I. These symmetry-breaking strong forces may be of the following kinds:

1. $SU(3)$-invariant central forces which couple the spins or the unitary-spins of two quarks. These separate states with different S or different $SU(3)$ label β.

2. $SU(3)$-invariant noncentral forces, of spin-orbit or tensor type. These interactions split states with the same S and L according to their total spin J.

3. $SU(3)$-breaking effects, which satisfy hypercharge and isospin conservation. These may be one-quark effects (distinguishing the singlet quark λ from the doublet quarks (p, n)) or q–q potentials (which may or may not involve noncentral terms).

We have no fundamental understanding of these $SU(6)$-breaking forces, nor even of the superstrong forces, except what we can learn from the regularities observed in the patterns of baryonic levels.

The $(1s)^3$ configuration is naturally expected to lie lowest in mass. It has permutation symmetry S and orbital angular momentum $L = 0+$. As mentioned above, the low-lying baryonic states with parity $+$ are the $(\frac{1}{2}+)$ octet B (typically $N(939)$) and the $(\frac{3}{2}+)$ decuplet D (typically $\Delta(1236)$, which are quite well separated from the next states with parity $+$ (excepting, perhaps, the $P_{11}(1470)$ state, whose quantum numbers (I, J, w) are the same as those for the nucleon). It is natural to identify these states as a 56-representation with $L = 0+$. Since both the space wavefunction $(1s)^3$

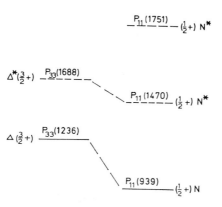

FIG. 1. The states P_{11} and P_{33} in Table I are displayed. The N and Δ states are the nucleonic components of the ground state $(56, 0+)$ supermultiplet. The first excited states $N^*(1470)$ and $\Delta^*(1688)$ are probably the nucleonic components of the first excited $(N = 2)$ $(56, 0+)$ supermultiplet. The interpretation of P_{11} (1751) is not known yet, but it appears quite likely that it may be a nucleonic component of a third $(56, 0+)$ super-multiplet, with excitation $N = 4$.

and the $SU(6)$ wavefunction for the 56-representation have permutation symmetry S, we conclude that the complete three-quark wavefunction is required to be totally symmetric with respect to permutations of the labels of the three quarks.

We note that the spin-spin splitting (or perhaps unitary spin-unitary spin splitting, or both) is measured typically by the difference $(\Delta\text{-}N) \approx 300$ MeV, and that the $SU(3)$-breaking interactions are measured typically by $(\Xi\text{-}N)/2 \approx 190$ MeV.

The $(1s)^2(1p)$ configuration is the first excited state. With harmonic shell-model states, the space wavefunction for this state has the form

$$\psi[(1s)^2(1p)] = \mathbf{r}_3 \exp[-\sigma(r_1{}^2 + r_2{}^2 + r_3{}^2)/2] \tag{2}$$

$$= \mathbf{r}_3 \, \psi[(1s)^3] \tag{3}$$

The symmetric part of this wavefunction is

$$\tfrac{1}{3}(\mathbf{r}_1 + \mathbf{r}_2 + \mathbf{r}_3)\psi[(1s)^3] \equiv 0 \tag{4}$$

since the center of mass of the system is at rest. The wavefunction $\psi[(1s)^2(1p)]$ clearly has no totally antisymmetric part. Therefore, the space wavefunction for this configuration necessarily has permutation symmetry M. Hence, in order that the complete three-quark wavefunction should have S symmetry, the only $SU(6)$-representation which can occur with the configuration $\psi[(1s)^2(1p)]$ (recalling that $M \times M = S + A + M$, whereas $M \times S = M$ and $M \times A = M$) is that with M symmetry, namely the 70-representation. The first excited supermultiplet is therefore predicted to consist of the states $(70, L = 1-)$.

From Table II, we see that the $SU(6)$-breaking interactions will separate this supermultiplet into the following nine unitary multiplets:

$$(70, L = 1-) \to (\{1\}, {}^2P_{1/2,3/2}) + (\{8\}, {}^2P_{1/2,3/2})$$
$$+ (\{10\}, {}^2P_{1/2,3/2}) + (\{8\}, {}^4P_{1/2,3/2,5/2}) \tag{5}$$

There is just one nucleonic candidate in Table I for each of these unitary multiplets between the lowest N^* with the negative parity $(D_{13}(1520))$ and the N^* with mass 1710 MeV. The next highest negative parity state in Table I is $D_{35}(1954)$, which is well separated in mass from this supermultiplet. The only two unitary singlets which are known to date have the spin-parity values required by the first term of (5), namely $Y_0^*(1405)$ with $(\tfrac{1}{2}-)$, and $Y_0^*(1520)$ with $(\tfrac{3}{2}-)$. The mass difference $[Y_0^*(1520) - Y_0^*(1405)] \approx 115$ MeV provides a convenient measure for the strength of the spin-orbit interaction in the q–q system. The second term of (5) corresponds to the states $D_{13}(1520)$ and $S_{11}(1535)$. The third term of (5) corresponds to the states $S_{31}(1640)$ and $D_{33}(1691)$. The three states which correspond to the last term of (5) are $S_{11}(1710)$, $D_{13}(1675)$ and $D_{15}(1680)$. These identifications are summarized in Fig. 2.

There could be some ambiguity about the identifications for the S_{11} and D_{13} states, especially as the spin-orbit interactions will generally mix the ^4P and ^2P states with the same J. The identifications given here are supported by independent arguments in two ways:

1. *The level widths.* Mitra and Ross[10] have made calculations of the partial widths for the s-wave and d-wave decays [to final states $\pi +$ $(56, L = 0+)$] and for the p-wave decays [to final states $\pi + (70, L = 1-)$] for all the members of the supermultiplet $(70, L = 1-)$. The final expressions contain a number of integrals over the space wavefunctions, which are treated as adjustable parameters, chosen to fit the empirical values for the partial widths. These calculations predict that the $^4P_{3/2}$ state should be strongly inelastic; the state $D_{13}(1675)$ has elasticity x only about 0.2, whereas $D_{13}(1520)$ has elasticity about 0.55. They predict also that the $^4P_{1/2}$ state should be very broad; $S_{11}(1710)$ has width about 300 MeV, compared with width 120 MeV for $S_{11}(1535)$. It could still be, of course, that there is substantial mixing between the ^4P and ^2P states, which could blur this argument appreciably. [cf. Note (i) added in proof.]

2. *The photoexcitations* $\gamma N \rightarrow N^*$. Moorhouse[11] has pointed out that the γ-transitions $^2S \rightarrow {}^4P$ are forbidden, whereas those for $^2S \rightarrow {}^2P$ are allowed. The transition $\gamma p \rightarrow N^*(1680, \frac{5}{2}-)^+$ is known to be very weak, in accord with this prediction. The two γ-transitions $\gamma p \rightarrow N^*(1520, \frac{3}{2}-)$ and $\gamma p \rightarrow N^*(1535, \frac{1}{2}-)$ are both well established (the latter because of its characteristic decay mode $N^*(1535, \frac{1}{2}-) \rightarrow \eta p$, so that it gives rise to a strong $\gamma p \rightarrow p\eta$ reaction near the $p\eta$ threshold), which provides a confirmation for their assignment to the ^2P configurations.

The pattern of spin-orbit splitting in the $(70, L = 1-)$ supermultiplet is not yet understood. For the ^2P states, the observed splittings are as follows:

	{1}	{8}	{10}
$M(\frac{3}{2}-) - M(\frac{1}{2}-)$	+115 MeV	−15 MeV	+50 MeV
F^2 (total unitary spin)	0	18	36

The assumption of a q–q spin-orbit interaction of the form

$$V_{SO} = (\sigma_1 + \sigma_2) \cdot L_{12}[1 + \lambda F(1) \cdot F(2)]v(r_{12}) \tag{6}$$

leads to a spin-orbit contribution to the mass of these states, given by the expression[12]

$$M_{SO} = (A + BF^2)\text{S.L.} \tag{7}$$

With this expression, the spin-orbit splitting in the $^2P(\{8\})$ states is required to be the average of those for $^2P(\{1\})$ and $^2P(\{10\})$, namely about $+82.5$ MeV, in contrast with the -15 MeV observed. Further, for the $^2P\,\Delta$ state obtained from the configuration $(1s)^2(1p)$, the expectation value of (6) vanishes, as Greenberg and Resnikoff[13] and Dalitz[14] have pointed out, in contrast with the observed splitting of 50 MeV. A spin-orbit coupling of tensor form would not help the situation, unless it were exceedingly strong, since a tensor interaction has zero expectation value for ^2P states.

Two of the ^4P states are identified here with the $S_{11}(1710)$ and $D_{15}(1680)$ resonances. These are both broad levels, with widths of order 200–300 MeV. In view of the uncertainties in the determination of resonance mass values, this mass splitting should be considered less than the limits of resolution. With $D_{13}(1675)$, the three ^4P states thus appear essentially degenerate. It appears that either (1) the spin-orbit interaction vanishes for octet states—this would essentially be consistent with the evidence for the $^2P(\{8\})$ states, where the splitting is also only -15 MeV—or (2) the spin-orbit interaction vanishes for $S = \frac{3}{2}$ states of the qqq system.

As we shall see, the information from higher supermultiplets suggests that the latter possibility is the case.

Here we should draw attention to the elaborate shell-model calculations carried out by Greenberg and Resnikoff,[13] who considered the $(1s)^3$ and $(1s)^2(1p)$ configurations together, including spin-orbit interactions and $SU(3)$-breaking potentials. They determined the potential parameters by using a least squares procedure, to fit the mass values for the $(\{8\}, \frac{1}{2}+)$ and $(\{10\}, \frac{3}{2}+)$ states and for seven well-established states of negative parity ($Y_0^*(1405)$, $Y_0^*(1520)$, $N^*(1520)$, $\Delta^*(1670)$, $N^*(1680)$, and $\Sigma^*(1770)$). The most controversial feature of their fit is the inclusion of a strong $SU(3)$-breaking spin-orbit force, which frequently gives rise to strong mixing between all the states with the same hypercharge and spin-parity. Their predictions for the N^* and Δ^* states do not fit the pattern of Fig. 2 very well. For the ^4P states, their calculation predicts 1744 MeV for $N^*(D_{13})$, and 1784 MeV for $N^*(S_{11})$, whereas these states are in fact essentially degenerate with $D_{15}(1680)$. For the ^2P states, the Δ^* states D_{33} and S_{31} are predicted to be degenerate, whereas in fact they are separated by 50 MeV. We shall not discuss the details of Greenberg and Resnikoff's interesting calculation here.

The $((1s)(1p)^2 + (1s)^2(1d) + (1s)^2(2s))$ configurations include six supermultiplets, given as follows by Greenberg[9,15]:

$SU(6)$ representation	56	56	70	70	20
$(L, w) =$	2+	0+	2+	0+	1+

The 20-representation comes entirely from the $(1s)(1p)^2$ configuration; the

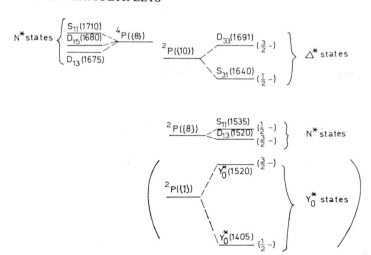

FIG. 2. The pattern of isospin multiplets comprising the $N = 1$ $(70, 1-)$ super-multiplet is displayed, by giving the nucleonic component established for each octet or decuplet, together with the Y_0^* singlet states.

others involve linear superpositions of the $(1s)(1p)^2$, the $(1s)^2(2d)$, and (for the case of $L = 0$), the $(1s)^2(2s)$ configurations.

As remarked earlier, we expect the 56-representation to lie lowest in each configuration. Here we consider first the $(56, L = 2+)$ supermultiplet, which we expect on rather general grounds to occur as a Regge rotational excitation of the $(56, L = 0+)$ ground configuration; the pattern of levels to be expected has already been given in Ref. 16. The $^2D(\{8\})$ states will give rise to N^* resonances in the P_{13} and F_{15} partial waves; the $^4D(\{10\})$ states will give rise to Δ^* resonances in the P_{31}, P_{33}, F_{35}, and F_{37} partial waves. We note that between the $P_{33}(1688)$ and $F_{37}(1950)$ resonances listed in Table I, there is just one candidate for each of these positive-parity resonance levels. The identifications made for the $(56, L = 2+)$ supermultiplet are shown on Fig. 3. The only additional positive-parity state listed in this mass range is another P_{11} level, at 1751 MeV, which we shall discuss below. For higher mass values, the only further positive parity state on Table I is $F_{17}(1983)$. The lowest supermultiplet which could give rise to this level is $(70, L = 2+)$ whose excited nucleon states are shown in Fig. 4; it is possible that this F_{17} level signals the presence of a $(70, L = 2+)$ supermultiplet in the mass region just above the limit of the well-explored mass range.

From the $(56, L = 2+)$ level pattern shown in Fig. 3, we see that the $S = \frac{1}{2}$ states again undergo a strong spin-orbit splitting, by about -170

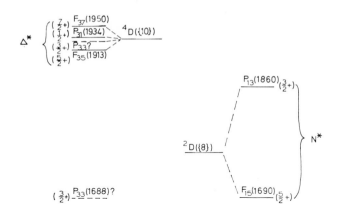

FIG. 3. The pattern of isospin multiplets comprising the $N = 2$ (56, 2+) super-multiplet is displayed, by specifying the N^* or Δ^* state belonging to each octet or de-cuplet, respectively. The only excited P_{33} state available in Table I is P_{33} (1688), which is drawn on this Figure. However, it is far from the other three $S = \frac{3}{2}$ levels and we conclude that there must be another P_{33} state, with mass near to 1930 MeV, still to be detected.

MeV, the state with the higher spin lying lower. The $S = \frac{3}{2}$ levels lie close together, near 1930 MeV, except for the $P_{33}(1688)$ level which is 250 MeV below the other three levels. This exceptional separation suggests that it is not correct to assign this $P_{33}(1688)$ level to the (56, $L = 2+$) supermultiplet. It appears more likely that $P_{33}(1688)$ is closely related with the $P_{11}(1470)$ level and that these two states belong to a further 56 supermultiplet; we discuss this possibility below in considering the possible interpretations for the P_{11} levels. The $S = \frac{3}{2}$ levels identified are all broad ($\Gamma = 200$–300 MeV) and lie in a mass range of only 40 MeV, so that these decuplet states P_{31}, F_{35}, and F_{37} may be considered as essentially degenerate. This is reminiscent of the (70, $L = 1-$) supermultiplet, where the $S = \frac{3}{2}$ octet states were found to be essentially degenerate. These two situations together suggest that the spin-orbit interaction for qqq systems vanishes (or is weak) for states with $S = \frac{3}{2}$. For the (56, $L = 2+$) supermultiplet, this leads to the *prediction* that there should exist another P_{33} level degenerate with the P_{31}, F_{35}, and F_{37} levels, i.e., close to the mass range 1910–1950 MeV. We note also that the (56, $L = 2+$) supermultiplet does not contain any Y_0^* singlet states; in fact apart from $Y_0^*(1405)$ and $Y_0^*(1520)$, the only Y_0^* singlet states which are to be expected below 1950 MeV are those which might be associated with $P_{11}(1750)$, according to the possibilities discussed below.

We have no independent checks on these assignments to the (56, $L = 2+$) supermultiplet. However, it is certainly striking that the simplest supermultiplet which can include both the long-established levels $F_{15}(1690)$ and $F_{37}(1950)$ should require the further existence of just those positive parity levels which have been observed in the mass range which they cover.

Higher N* and Δ* Levels. We have already mentioned the level $F_{17}(1983)$. Its assignment to the only available supermultiplet, the $(70, L = 2+)$ shown on Fig. 4, requires the existence of related resonance levels in the F_{15} (twice), P_{13} (twice), P_{11}, F_{35} and P_{33} partial waves, together with Y_0^* singlet states with spin-parity $(\frac{5}{2}+)$ and $(\frac{3}{2}+)$.

The other high-mass levels listed in Table I have negative parity, $D_{35}(1954)$, $D_{13}(2057)$, and $G_{17}(2200)$. The most immediate possibility is that they belong to supermultiplets with three-quantum excitation, coming from the configurations $(1p)^3$, $(1s)(1p)(1d)$, $(1s)(1p)(2s)$, $(1s)^2(2p)$, and $(1s)^2(1f)$, which allow L values of 1, 2, and 3, with negative parity. According to Greenberg[9] and to Karl and Obryk[15], the following supermultiplets are allowed from these configurations:

$SU(6)$ representation	56	56	70	70	70	20	20
(L, w)	$3-$	$1-$	$3-$	$2-$	$1-$	$3-$	$1-$

The $(56, L = 3-)$ supermultiplet includes N^* states D_{15} and G_{17}, and Δ^* states D_{33}, D_{35}, G_{35}, and G_{37}. Two of the three observed negative parity levels could be fitted into this supermultiplet. It is not obvious that this will necessarily be the lowest supermultiplet for three-quantum excitation. The Regge-rotational excitation of the only one-quantum supermultiplet $(70, L = 1-)$ will be the supermultiplet $(70, L = 3-)$. The

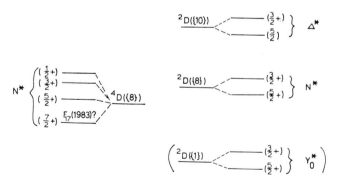

FIG. 4. The pattern of isospin multiplets for the $N = 2$ $(70, 2+)$ supermultiplet is sketched. There is only one state in Table I, namely $F_{17}(1983)$, which requires this identification at present. There is no theoretical reason to expect this $(70, L = 2+)$ supermultiplet to lie at such low mass values, comparable with the mean mass for the $(56, L = 2+)$ supermultiplet.

lower supermultiplet which would be expected to correspond to the $(56, L = 3-)$ supermultiplet is the supermultiplet $(56, L = 1-)$. However, this is a spurious state of the configuration $(1s)^2(1p)$ and therefore does not exist; we may expect this fact to be reflected in some corresponding constraints on the $(56, L = 3-)$ state, which might prevent it from being the lowest supermultiplet for this configuration. The $(70, L = 3-)$ supermultiplet can accomodate all three negative parity levels, since it includes the N^* states D_{13}, D_{15}(twice), G_{17}(twice), and G_{19}, and the Δ^* states D_{35} and G_{37}, together with Y_0^* singlet states $(\frac{5}{2}-)$ and $(\frac{7}{2}-)$. There exists a strong candidate[8] for the last of these states, the resonance $Y_0^*(2100)$, which has no known N^* counterparts and which is generally believed to have spin-parity $(\frac{7}{2}-)$.

Two of these states $[D_{13}(2057)$ and $D_{35}(1954)]$ could belong to the supermultiplet $(56, L = 1-)$, but this supermultiplet does not include partial waves as high as G_{17}. The supermultiplet $(70, L = 2-)$ could accommodate all these levels, since it requires N^* states G_{17}, D_{15}(twice), D_{13} (twice), and S_{11}, Δ^* states D_{33} and D_{35}, and Y_0^* states with spin-parity $(\frac{5}{2}-)$ and $(\frac{3}{2}-)$. We note that this supermultiplet does not include the (probably) singlet state $Y_0^*(2100)$.

The only three-quantum supermultiplet that can accommodate all four of the negative-parity baryonic states known near the 2000–2200 MeV range is $(70, L = 3-)$. This leads to the expectation of a considerable number of further N^* and Δ^* states with negative parity in this mass region—in particular, the systematics of the $L = 1-$ and $2+$ supermultiplets then suggest that there should exist, D_{13}, D_{15}, and G_{19} resonance levels essentially degenerate with $G_{17}(2200)$[18]—and their observation will provide an important test for the identifications proposed for the levels existing in this mass region.

The Outstanding Positive Parity Levels. The nature of the level $P_{11}(1470)$ has been a mystery for a number of years. Now the phase-shift analysis summarized on Table I provides us with a further P_{11} level at 1751 MeV.

With the configurations $N < 4$ (for N-quantum excitations), there are a number of possibilities for the interpretation of these P_{11} levels:

1. The supermultiplet $(56, L = 0+)$ for $N = 2$. This corresponds to an N^* resonance in the P_{11} state, together with a Δ^* resonance in the P_{33} state. The most natural supposition would be that these states should lie roughly at the same mass level as holds for the other $N = 2$ states, namely at about 1900 MeV. We shall adopt the notation $(56, L = 0+)^*$ below for this supermultiplet.

2. The supermultiplet $(20, L = 1+)$ from $N = 2$ states. This supermultiplet consists of two octets, with spin-parity $(\frac{3}{2}+)$ and $(\frac{1}{2}+)$, and three

Y_0^* triplets, with spin-parity $(\frac{1}{2}+)$, $(\frac{3}{2}+)$ and $(\frac{5}{2}+)$. Hence only two N^* states are expected, in the waves P_{11} and P_{13}.

3. The supermultiplet $(70, L = 2+)$, from the $N = 2$ excitations, as shown in Fig. 4. The ^4D states are octet, and give rise to N^* resonances in the partial waves P_{11}, P_{13}, F_{15}, and F_{17}, as mentioned above. The ^2D states lead to P_{13} and F_{15} N^* resonances, to P_{33} and F_{35} Δ^* resonances, and to Y_0^* resonances with spin-parity $(\frac{3}{2}+)$ and $(\frac{5}{2}+)$.

4. The supermultiplet $(70, L = 0+)$, from the $N = 2$ excitations. Its ^4S and ^2S states lead to N^* resonances in the partial waves P_{13} and P_{11}, respectively, and to a Δ^* resonance P_{31}. One Y_0^* state is also required, with spin-parity $(\frac{1}{2}+)$.

Of these four possibilities, the most plausible assignment for $P_{11}(1470)$ is the $(56, L = 0+)^*$ supermultiplet. This is certainly the most economical assignment. This supermultiplet parallels the ground state 56-supermultiplet, and the only other excited nucleon state it contains is a Δ^* state, again with spin-parity $(\frac{3}{2}+)$. The observed level $P_{33}(1688)$ falls naturally into this interpretation, especially as its mass relationship with $P_{11}(1470)$ is exceedingly similar to that of $\Delta(1236)$ with the corresponding $N(939)$. the (mass)^2 difference $(P_{33}^2 - P_{11}^2)$ having the values 0.69 (GeV)^2 for the upper pair, and 0.65(GeV)^2 for the lower pair. These states $P_{11}(1470)$ and $P_{33}(1688)$ do lie rather low in mass, relative to the other $N = 2$ excitation $(56, L = 2+)$, in fact at mass comparable with the mean mass of the nucleonic components of the $N = 1$ excitation $(70, L = 1-)$. This may be due to the fact that the $N = 2$ state $(56, L = 0+)$ involves only s-wave qq interactions, its space wavefunction being $(r_{12}^2 + 3r_3^2)\psi[(1s)^3]$, where r_1, r_2, and r_3 denote the quark position vectors relative to their center of mass.

The only other possibility of interest would be the assignment of $P_{11}(1470)$ to $(20, L = 1+)$. This would require also the existence of a companion P_{13} level, for which the only available candidate is $P_{13}(1863)$; this identification would involve the assumption of a spin-orbit splitting by about 400 MeV, which is very much larger than the other spin-orbit splittings known. It would be surprising for this configuration to lie so low in mass. Its space wavefunction has the form $(r_1 \times r_2 + r_2 \times r_3 + r_3 \times r_1)\psi[(1s)^3]$, which involves relative p-wave motion between each pair of quarks. We would expect the qq attractions to be much less effective in such a configuration, relative to those where the s-wave qq attraction can be effective [as in the configuration $(56, L = 2+)$, where the internal orbital angular momenta are $(l_{12}, l_3) = (2, 0)$ and $(0, 2)$]. This remark is entirely equivalent to our earlier remark that, with simple qq attractions, we expect 56 states to lie lower than 70 and 20 states in the same shell-model configuration. The supermultiplet $(20, L = 1+)$ could lie low in mass only if there were a very strong space-exchange component in the qq interaction, giving added attraction to the p-state interaction, so that it became much stronger than the s-state qq interaction. This supermultiplet would also require three singlet Y_0^* states, for which there is no evidence. On the other hand, the mass range 1520–1660 MeV has been relatively little explored for Y_0^* systems, and it cannot be excluded at present that there might exist such Y_0^* states, still to be found.

The last two possibilities, $(70, L = 2+)$ and $(70, L = 0+)$, both include a P_{11} state. The former includes both P_{11} and P_{33} states, but also requires a number of other nucleonic states in the mass range already well explored, for which there are no indications at present. The latter requires P_{13} and P_{31} levels, for which there is no evidence in the relevant mass range, but does not provide any place for $P_{33}(1688)$. These M representations also do not have any particularly attractive features, from the standpoint of dynamics.

Adopting the identification of $P_{11}(1470)$ and $P_{33}(1688)$ as members of the $(56, L = 0+)^*$ supermultiplet, we turn to the remaining state $P_{11}(1750)$. This must belong to the $N = 2$ excitations, or to the $N = 4$ excitations. There are two $(56, L = 0+)$ states in the $N = 4$ configurations[17]; however, it would be surprising indeed to find such $N = 4$ excitations lying so low in mass. The $N = 2$ supermultiplet $(70, L = 2+)$ has the disadvantage that it requires five other nucleonic states not far from $P_{11}(1750)$ in mass value. We have already suggested that $F_{17}(1983)$ must belong to this super-multiplet (the only one available for it, with $N < 4$). This would require a spin-orbit splitting of 200 MeV between these P_{11} and F_{17} states. This would be in conflict with the empirical rule that the spin-orbit splittings are weak for $S = \frac{3}{2}$ states (but we have no theoretical understanding of this rule, anyway), and would also require that the corresponding P_{13} and F_{15} states should lie close to the mass range 1750–1980 MeV. The state $P_{13}(1863)$ could belong here, although it has already been identified with the supermultiplet $(56, L = 2+)$; there is no F_{15} (nor F_{35}) candidate in this mass range. The most likely (and most economical) $N = 2$ assignment for $P_{11}(1750)$ is to the supermultiplet $(70, L = 0+)$; there are candidates [$P_{13}(1863)$ and $P_{31}(1934)$] for the accompanying nucleonic states, although we have already assigned these to the supermultiplet $(56, L = 2+)$. The only other possibility is that $P_{11}(1750)$ belongs to the supermultiplet $(20, L = 1+)$, which would require a companion state P_{13} [which could be the level $P_{13}(1863)$].

We conclude that all of the $N = 2$ assignments available for $P_{11}(1750)$ require the existence of at least one other nucleonic level which has positive parity and lies close in mass to the level $P_{11}(1750)$, and of at least one Y_0^* singlet with positive parity in this neighborhood. However, it is not excluded that $P_{11}(1750)$ belongs to $(56, 0+)$ and $N = 4$; this would not require any corresponding Y_0^* state.

Discussion. It appears remarkable that the nucleonic resonance levels observed up to 2200 MeV should appear in groupings which correspond in turn to the supermultiplets $(56, L = 0+)$ for excitation $N = 0$, $(70, L = 1-)$ for $N = 1$, $(56, L = 2+)$ and (probably) $(56, L = 0+)$ for $N = 2$. These supermultiplets are just those predicted for the case of shell model wave-functions with excitation quanta $N = 0$ and 1, and those expected to lie lowest in mass for $N = 2$. There are several outstanding levels (F_{17}, G_{17}, D_{35}, D_{13}, and P_{11}) at the upper end of the mass range explored, which can be plausibly identified with two or more of the supermultiplets expected to lie just above this mass range. There is also one nucleonic level which has been predicted to fill out these $N = 2$ supermultiplets but which has not yet been detected; this is the state P_{33}, predicted to be close to 1930 MeV. The $P_{11}(1751)$ state remains a puzzle. The most plausible $N = 2$ supermultiplet to fit this state is $(70, L = 0+)$, but the accompanying nucleonic states P_{13} and P_{31} have not been seen or have been wrongly identified; in either case, completion of these $N = 2$ supermultiplets requires additional states P_{13} and P_{31}. On the other hand, it is an exceedingly attractive and quite plausible notion that the states $P_{11}(939)$, $P_{11}(1470)$, $P_{11}(1750), \ldots$ may form a series of related 56 excitations, corresponding essentially to radial excitations In this case, $P_{11}(1750)$ would have to be assigned to an $N = 4$ configuration (built from $(1s)(2s)^2$ and $(1s)^2(3s)$, together with those configurations which must be taken together with these to eliminate spurious states). This would require a second missing P_{33} level to lie in the mass region 1900–2000 MeV. The detection of further positive parity levels (perhaps very strongly inelastic) in the mass range already explored is to be expected, and would be very instructive in testing the level patterns already proposed and in deciding between the various possibilities for interpretation of $P_{11}(1750)$. At present, there are no quanti-tative predictions available on the decay modes and level widths for the individual isospin multiplets belonging to the higher supermultiplets. When these are available, it will be important to test these supermultiplet assign-ments quantitatively. [cf. Notes added in Proof, at end.]

The configurations up to $N = 3$, and the supermultiplets they give rise to are tabulated in Table III, following Karl and Obryk.[17] Those supermultiplets which correspond to the observed levels up to about 2000 MeV are marked with a dagger. We note that these are but a small fraction of the total number of supermultiplets, and we conclude that for the next mass region to be explored, say 2000–3000 MeV, there are likely to exist a very large number of unitary multiplets. Most of these resonant states may be expected to be rather broad, and highly inelastic, so that one

TABLE III

The Shell-Model Configurations for N Quanta of Excitation

Excitation Quanta N	Configurations	Supermultiplets
$N = 0$	$(1s)^3$	$(56, 0+)^†$
$N = 1$	$(1s)^2(1p)$	$(70, 1-)^†$
$N = 2$	$(1s)(p)^2$	$(56, 2+),^†$ $(56, 0+),^†$ $(70, 2+),$
	$(1s)^2(1d)$	$(70, 0+)$ and $(20, 1+)$
	$(1s)^2(2s)$	
$N = 3$	$(1s)^2(1f)$	$(56, 3-)$ $(56, 1-),$ $(70, 3-)$ $(70, 2-),$
	$(1s)^2(2p)$	$(70, 1-)$ twice, $(20, 3-),$
	$(1s)(1p)(1d)$	and $(20, 1-)$
	$(1s)(1p)(2s)$	
	$(1p)^3$	

Listed for $N \leq 3$ together with the supermultiplets (α, Lw) to which they can contribute.[17]

The supermultiplets marked with a dagger are those for which there is a detailed identification in the empirical data. For $N \geq 3$, the supermultiplets listed can occur more than once [as noted on the table for $(70, 1-)$]. Spurious states have been omitted wherever they occur.

would require very accurate and detailed observations on the πN interactions before one could hope to detect and establish them. The next step to be taken forward in the study of baryonic resonances may well prove to be rather difficult.

In the above considerations, it has been assumed without comment that the complete three-quark wavefunctions for baryonic states should be totally symmetric in the labels of the three quarks, as was required by the shell-model wavefunctions, taken together with our empirical knowledge of the low-lying baryonic states with positive parity. However, quarks have spin $\frac{1}{2}$, and the spin-statistics theorem tells us that Bose statistics is not permitted for particles with half-integer spin. One way out of this dilemma, as pointed out rather early by Greenberg,[9] is the supposition that quarks obey parafermi statistics of order $p = 3$.

Parastatistics? This possibility may be understood most simply in the following way, using the representation for the quark field operator, as first proposed by Green[19] (who had proposed earlier in the same paper, the set of commutation relations which form the abstract definition of parastatistics) and shown quite generally by Greenberg and Messiah[20] to be equivalent with the field-theoretic formulation of parastatistics:

$$q_\alpha(n) = \sum_{i=1}^{p} q_\alpha^i(n) \tag{8}$$

Here n denotes the label identifying a particular quark, and the integer p is known as the order of the parafermi statistics. For quarks, the choice $p = 3$ is required in order that a proton should have statistical weight unity. The three-quark states q_α^1, q_α^2, and q_α^3, for given α, represent different but physically indistinguishable particles. This is ensured by the requirement that, in the interaction Lagrangians, the quark fields enter only in the combination (8); the physical interactions of two quarks q_α^i with the same suffix α are exactly the same, independent of the superfix i. The essential feature of this scheme is that the commutation relations between two quark fields q_α^i and q_β^j depend on the relationship of i and j. For $i = j$, the two fields obey the anticommutation relations characteristic of Fermi statistics; for $i \neq j$, the two fields obey the commutation relations characteristic of Bose statistics.

It is this last feature which allows the three-quark wavefunctions for the baryon states to have the symmetry appropriate to Bose statistics; these states can be built up using three quarks of different type (that is q_α^1, q_β^2, q_γ^3, for example). We should note that it is equally possible to build up three-quark states which have the symmetry appropriate to Fermi statistics, or even mixed symmetry. Essentially, the situation with parastatistics may be described as the introduction of a new discrete quantum number i, with the eigenstates $\chi^i(n)$ for the quark labeled n, where i can take the values 1, 2, 3. For a three-quark system, wavefunctions associated with this new degree of freedom can be constructed from the basic eigenstates $\chi^i(1)\chi^j(2)\chi^k(3)$ to have permutation symmetry S, M, or A. The complete wavefunction, now including space variables, spin, unitary-spin, and the new discrete variable i, is still required to be antisymmetric in the particle labels; however, the wave-function ψ for space, spin, and unitary-spin variables can now have permutation symmetry A, M, or S. Which symmetry the wavefunction ψ (including space, spin, and unitary spin variables) will have will depend on the detailed nature of the q–q interaction. It has generally been assumed (tacitly) in the literature that the permutation symmetry appropriate for all the low-lying states is the same as that required for the ground state; it is to be expected that there will generally exist high-lying states with symmetry different from that for the ground state. If the q–q superstrong potential has no dependence on spin and unitary spin, then, for example, the following three $(1s)^3$ states would have the same energy:

$$\chi_A \phi(56) \psi_{sp}.[(1s)^3] \tag{9a}$$

$$\chi_S \phi(20) \psi_{sp}.[(1s)^3] \tag{9b}$$

$$[\chi_1 \phi_2(70) - \chi_2 \phi_1(70)] \psi_{sp}[(1s)^3] \tag{9c}$$

where the functions ϕ refer to spin and unitary spin, the functions χ refer to the new internal variable i, and the sets (ϕ_1, ϕ_2) and (χ_1, χ_2) each form a standard basis for a mixed representation for the permutation group on three objects. However, in general, these three states (9) will be separated by energy differences typical of the superstrong interactions (which are measured by the quark mass $2M_q \gtrsim 10$ GeV), in consequence of spin and unitary-spin dependence of the q–q interaction, consistent with $SU(6)$ symmetry.

The internal variable i cannot be detected by direct observation, since the quark interactions in this parastatistics scheme are always such that they are precisely the same for each value of i. On the other hand, the existence of this variable i could be deduced indirectly (for example, by testing detailed balancing for a reaction such as $\pi + \pi \rightleftharpoons q + \bar{q}$) in situations which depend on the statistical weight for the quark (since this has the value $p = 3$ with our assumptions concerning the quark in this Section).

The Han-Nambu Model[21] should be mentioned here in view of its superficial parallelism with the parastatistics assumption. This model envisages three quark triplets, $(q_\alpha{}^i)$ for $i = 1, 2, 3$. These quarks all have integral charge; the p-quark for the triplet $q_\alpha{}^i$ has charge $+1$, whereas the p-quark for each of the triplets $q_\alpha{}^2$ and $q_\alpha{}^2$ has zero charge. The three quark triplets $(q_\alpha{}^i)$ are here different and physically distinguishable (for interactions less than superstrong); for example, as we have just said, the quarks p^1 and p^2 have different charge, hence different electromagnetic coupling. Any pair $(q_\alpha{}^i$ and $q_\beta{}^j)$ of quark fields obey the anticommutation relations appropriate to Fermi statistics.

Han and Nambu suppose that the superstrong forces obey an $SU(3)'$-symmetry with respect to the *upper suffix* for the quark. Next, they suppose that the superstrong forces are such as to favor strongly $SU(3)'$-singlet states. For a three-quark system $q_\alpha{}^i(1)q_\beta{}^j(2)q_\gamma{}^k(3)$, the low-lying baryon states will then correspond to the $SU(3)'$-singlet states,

$$\sum_{i,j,k} \varepsilon_{ijk} \, q_\alpha{}^i(1) q_\beta{}^j(2) q_\gamma{}^k(3) \tag{10}$$

The antisymmetric character of the ε_{ijk} factor means that the corresponding wavefunction $\psi(1, 2, 3)$ for space, spin, and unitary-spin can be symmetrical in the labels 1, 2, 3, as required by the empirical situation.

This scheme explicitly involves new additive quantum numbers for the superstrong interactions. These $SU(3)'$ quantum numbers do not come into play for interactions involving the low lying states, since these are all $SU(3)'$-singlets. At sufficiently high excitation energy, there would exist states with nonzero $SU(3)'$ quantum numbers, however, there is no empirical evidence for the presence of such additional quantum numbers at present.

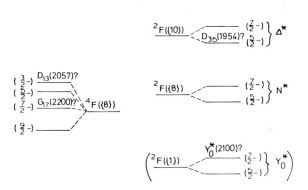

FIG. 5. The pattern of isospin multiplets for the $N = 3$ $(70, L = 3-)$ super-multiplet is displayed, by giving the nucleonic component established for each octet or decuplet, together with the Y_0^* singlet states. There are four states known which appear to be candidates for assignment to this supermultiplet, including the state $Y_0^*(2100)$ whose singlet character appears plausible but unproven to date.

Notes added in proof

(*i*) Dr. A. N. Mitra has pointed out that we have misinterpreted the tables given in Ref. 10) for the S_{11} states. It is the $^2P_{1/2}$ state for which a large πN width is predicted rather than the $^4P_{1/2}$ state. From dynamical considerations (the systematics of the spin-orbit term, as discussed above, or from the detailed calculations of Ref. 13), the identification of $S_{11}(1710)$ and $S_{11}(1535)$ with the states $^4P_{1/2}$ and $^2P_{1/2}$, respectively, is very plausible. However, the πN width then predicted for $S_{11}(1535)$ is $\Gamma_{\pi N} = 300$ MeV, compared with the empirical value of about 40 MeV; for $S_{11}(1710)$, the predicted πN width is 130 MeV, compared with the empirical value of about 240 MeV.

Faiman and Hendry[22]) have discussed the πN widths for the N^* an Δ^* states up to the $N = 2$ excitations, on the basis of the harmonic shell model discussed above. They have included mixing between levels with the same J within a supermultiplet, in a phenomenological spirit. The mixing angle θ for the $P_{1/2}$ configurations must have value $35°$ or $90°$ in order to fit the observed S_{11} widths: the latter solution just corresponds to an interchange between the assignments given above for the S_{11} states, and the dynamical considerations make this a rather unpalatable solution. The dynamical calculations of Greenberg and Resnikoff[13]) and of Divgi and Greenberg[23]) lead to mixing angle $\theta = 0°$; in fact they find essentially no mixing for any N^*, Δ^*, or Ξ^* states, whereas the mixing between Y_0^* and Y_1^* states is frequently very strong.

Faiman and Hendry[24]) have also calculated the widths $\Gamma_{N\gamma}$ and $\Gamma_{\Delta\gamma}$ for the nucleonic states according to their harmonic shell model. For the $S_{11}(1535)$ state, $\theta = 90°$ does not give an acceptable radiative width (in fact, γN decay is forbidden from all ^4P states), whereas $\theta = 35°$ (or any other value between $\pm 45°$) does predict an acceptable $\Gamma_{N\gamma}$.

(ii) For the Roper resonance $P_{11}(1470)$, Faiman and Hendry[22]) have calculated $\Gamma_{\pi N}$ for the various supermultiplet assignments possible. Only the $(56, L = 0+)^*$ assignment leads to a reasonably acceptable value, $\Gamma_{\pi N} = 50$ MeV compared with the empirical value of 140 MeV, the corresponding predictions for the $(70, L = 0+)$ and $(70, L = 2+)$ assignments being 16 MeV and 2 MeV, respectively. For the corresponding level $P_{33}(1688)$, $\Gamma_{\pi N}$ is then predicted to be about 50 MeV, to be compared with the empirical value about 30 MeV. Their calculation[24]) of the $P_{11}(1470)$ radiative width $\Gamma_{N\gamma}$ gives 0.13 MeV, and there is no doubt that $P_{11}(1470)$ photoexcitation has been observed to occur quite strongly, through observations on the channel $p\pi^+\pi^-$.

(iii) Greenberg and Nelson[26]) have pointed out that it may not be necessary to assume that the lowest baryonic supermultiplets are $SU(3)'$-singlets. They adopt an $[SU(3)' \times SU(3)'']$ scheme; the $SU(3)$ classification does not correspond to either $SU(3)'$ or $SU(3)''$, but to a combination of them. Including spin variables, they associate the $(1s)^3$ configuration with the 816-representation of $SU(18)$, the baryonic $SU(6)$ 56-supermultiplet being a submultiplet of this larger representation. This grand supermultiplet consists of states with spin-parity $(\frac{1}{2}+)$ and $(\frac{3}{2}+)$, and includes $Y = 2$ resonance states, of types both Z_0 and Z_1.

References

1. A. Donnachie, R. Kirsopp, and C. Lovelace, *Phys. Letters*, **26B**, 161 (1968).
2. C. Lovelace, Proc. Heidelberg Intl. Conf. on Elementary Particles, North-Holland, 1968 p. 79.
3. C. Lovelace, this volume.
4. A. Donnachie and C. Lovelace, private communication (Jan. 1968).
5. R. Cool, G. Giacomelli, T. Kycia, B. Leontic, K. Li, and A. Lundby, *Phys. Rev. Letters*, **17**, 102 (1966).
6. D. Bugg, R. Gilmore, K. Knight, D. Salter, G. Stafford, E. Wilson, J. Davies, J. Dowell, P. Hattersley, R. Homer, A. O'Dell, A. Carter, R. Tapper, and K. Riley, *Phys. Rev.*, **168**, 1466 (1968).
7. R. Abrams, R. Cool, G. Giacomelli, T. Kycia, B. Leontic, K. Li, and D. Michael, *Phys. Rev. Letters*, **19**, 259 (1967).
8. A. H. Rosenfeld, N. Barash-Schmidt, A. Barbaro-Galtieri, L. Price, P. Söding, C. Wohl, M. Roos, and W. Willis, "Data on Particles and Resonant States," Lawrence Radiation Lab. Rept. UCRL 8030-Pt.1 Rev., January, 1969.
9. O. W. Greenberg, *Phys. Rev. Letters*, **13**, 598 (1964).
10. A. N. Mitra and M. Ross, *Phys. Rev.*, **158**, 1630 (1967).
11. G. Moorhouse, *Phys, Rev. Letters*, **16**, 771 (1966).
12. R. H. Dalitz, *Proc. 13th Intern. Conf. on High Energy Physics*, Univ. of California Press, Berkeley, California, 1967, p. 215.
13. O. W. Greenberg and M. Resnikoff, *Phys. Rev.* **163**, 1844 (1967).
14. R. Dalitz, "Hadron Spectroscopy," *Proc. 2nd Hawaii Topical Conf. in Particles*, (Univ. Hawaii Press), 1968, p. 325.

15. Greenberg's paper (Ref. 9) includes also a supermultiplet $(70, 1 = 1 +)$. E. Obryk and G. Karl have pointed out that this is a spurious state.[17] This is also the case for the supermultiplet $(20, L = 0-)$ listed by Greenberg for the three quantum excited states.
16. R. Dalitz, *Proc. Oxford Intern. Conf. Elementary Particles*, Rutherford High Energy Laboratory, Chilton, January, 1966, p. 157.
17. G. Karl and E. Obryk, *Nucl. Phys.* **B8**, 609 (1968). These authors have systematically constructed a list of all the allowed shell-model harmonic-oscillator states for all configurations of the qqq system, up to excitations $N = 8$. The entries given in Table III have been taken from their paper. I would like to acknowledge here some fruitful discussions with them about the details of these configurations and their systematic construction.
18. Note that the scattering data used in Refs. 1–4 does not extend as far as 2200 MeV. This work concerns data which lies far out on the lower wing of N^* (2200), but the G_{17} phase shifts are consistent with the existence of this resonance (which has been assigned to the G_{17} partial wave from other considerations[8]) beyond the region of the data. However, this does mean that other N^* resonances in the vicinity of 2000 MeV might well have escaped detection in this phase-shift analysis, especially if they are narrower or less elastic than the G_{17} level.
19. H. S. Green, *Phys. Rev.*, **90**, 270 (1953).
20. O. Greenberg and A. Messiah, *Phys. Rev.*, **138**, B1155 (1965).
21. M. Han and Y. Nambu, *Phys. Rev.*, **139**, B1006 (1965).
22. D. Faiman and A. W. Hendry, *Phys. Rev.*, **173**, 1720 (1968).
23. D. R. Divgi and O. W. Greenberg, *Phys. Rev.*, **175**, 2024 (1968).
24. D. R. Divgi, *Phys. Rev.*, **175**, 2027 (1968).
25. D. Faiman and A. W. Hendry, "Electromagnetic Decays of Baryon Resonances in the Harmonic Oscillator Model." *Phys. Rev.*, in press (1968).
26. O. W. Greenberg and C. A. Nelson, *Phys. Rev. Letters* **20**, 604 (1968).

Remarks on Low-Energy πN Scattering*

J. J. SAKURAI

*The Enrico Fermi Institute for Nuclear Studies
and the Department of Physics,
The University of Chicago, Chicago, Illinois*

We have heard a great deal about recent exciting developments in *high*-energy πN scattering. Yet when the subject of the πN interaction is reviewed using a time scale of the order of ten to fifteen years, the major theoretical progress appears to center around our recent successful understanding of the *low*-energy parameters. Specifically I shall restrict myself to just two experimental numbers, the symmetric and the antisymmetric part of the *s*-wave scattering lengths

$$a_{1/2} + 2a_{3/2} = (0.00 \pm 0.01)/m_\pi$$

$$a_{1/2} - a_{3/2} = (0.27 \pm 0.01)/m_\pi$$

To put these numbers in perspective, I may mention that if the scattering lengths were due to the nucleon pole alone, we would get

$$a_{1/2} + 2a_{3/2}|_{N\,pole} \approx -3\,\frac{(G^2_{\pi NN}/4\pi)}{m_N} \approx -6/m_\pi$$

$$\left.\frac{a_{1/2} + 2a_{3/2}}{a_{1/2} - a_{3/2}}\right|_{N\,pole} \approx \frac{2m_N}{m_\pi} \approx 14$$

The mysteries of *s*-wave πN scattering are then twofold:
  1. Why is the symmetric part smaller than is expected from the nucleon pole contribution by two orders of magnitude?†
  2. Why does the antisymmetric part dominate?

For the benefit of young readers I may emphasize that these questions are very old. In fact they existed even when I was a graduate student, many, many years ago. Having learned about the Feynman techniques, I started calculating the πN scattering amplitude in the γ_5 coupling theory.

* This work supported in part by U.S. Atomic Energy Commission.
  † We have learned from Lovelace (see his paper in this volume) that $a_{1/2} + 2a_{3/2}$ may be as large as $0.06/m_\pi$. This is still far from $-6/m_\pi$; note $(6/0.06)^2 = 10,000$.

The Born term in the γ_5 coupling theory is identical to what is now referred to as the nucleon pole contribution and therefore unsatisfactory. Next I used the gradient coupling theory and found out that the results were much more satisfactory because of the very small s-wave scattering lengths it predicts. Yet I was told by Professor Bethe that the gradient coupling theory should not be taken seriously because it is not renormalizable. Around that time dispersion theory just started becoming fashionable; so I asked whether one could distinguish the unsatisfactory γ_5 coupling theory from the satisfactory gradient coupling theory on the basis of this new all-powerful formalism. The answer I got from an expert was that this is not possible because the difference appears only in the subtraction constant, the two coupling schemes giving rise to the same pole contribution. Discouraged by this kind of answer, I decided to work on some other problems and wrote a doctoral thesis on weak interactions.

While I was working on the weak interactions, I was impressed by the roles of conserved currents and universality in particle physics. It therefore appeared natural, at least to me, to construct a theory of *strong* interactions based on conserved currents and universality. In this way I was led to predict vector mesons coupled universally to conserved currents. I then came to realize that there is nothing sacred about the γ_5 coupling since a theory in which the vector mesons are fundamental appeared to make just as much sense as a theory in which the pion is fundamental. In any case I preferred the gradient coupling because the source of the pion coupled through its gradient can be related to the axial vector current that appears in the semileptonic weak interactions. So, even though both the γ_5 theory and the gradient coupling *theory* are probably wrong, I argued, the gradient coupling *prescription* may still approximate reality.

If we once accept the gradient coupling prescription, we can immediately explain why the s-wave scattering lengths are small, but this still does not tell us why $a_{1/2} - a_{3/2}$ is much bigger than $a_{1/2} + 2a_{3/2}$. Fortunately one of the vector mesons I proposed is supposed to be coupled universally to the isospin and can be exchanged between the pion and the nucleon. This kind of reasoning enables us to write down a scattering-length formula for a pion on any target A

$$a_T = -[2\mu_\pi(f_\rho^2/4\pi)]/m_\rho^2 \mathbf{T}_\pi \cdot \mathbf{T}_A$$

μ_π, reduced pion mass

in Born approximation. The sign is such that we get repulsion in the $T = \frac{3}{2}$ state (isospins parallel) and attraction in the $T = \frac{1}{2}$ state (isospins anti-parallel) in agreement with experiment as far as the πN interaction is concerned. Later on, when the mass and the width of the ρ meson became known, I realized that the magnitude predicted by this formula is also right to an accuracy of about 20%.

In any case, by the beginning of 1960 I was convinced that the correct way to approach low-energy πN scattering was to use the gradient coupling to estimate the nucleon contribution and, in addition, take into account the exchange of the ρ meson assumed to be coupled universally to the isospin. Most experts at that time thought I was insane presumably because I did not use the then-fashionable language of double-dispersion relations to formulate my ideas.

In the past few years it has become increasingly evident that this conjecture is not complete nonsense. In retrospect the 1962 work of Nambu and co-workers deserves special attention. Using chirlity conservation, they obtained a concise formula for "soft-pion bremsstrahlung," which may be written diagrammatically as follows:

In this diagram c_π is the pion decay constant which may be set equal to $-(m_N/G_{\pi NN})(g_A/g_V)$ by virtue of the Goldberger-Treiman relation, and the symbol $\times$ stands for an interaction with an axial spurion. This interesting work of Nambu did not receive much attention partly because it was difficult to test the idea in pion-production experiments in a clean manner.

A couple of years later, however, Adler applied an essentially equivalent formalism to an even simpler "process." He focused his attention on the problem of attaching a soft-pion line to the basic πN vertex to obtain the πN scattering amplitude where either the incident or the outgoing pion has zero four momentum. Diagrammatically:

In this way he derived the now famous consistency condition

$$\lim_{q \to 0} A^{(+)} = G^2_{\pi NN}/m_N$$

(where I have used the standard CGLN decomposition for the invariant matrix $-A + i\gamma \cdot QB$).

In my opinion the true physical significance of this remarkable condition is almost completely obscured in the literature. The simplest way to understand this result goes as follows. As I remarked earlier, when I was a graduate student, I learned from a dispersion theory expert that the difference between the γ^5 coupling and the gradient coupling prescription lies in the subtraction constant. What the Adler condition tells us is that the subtraction constant must be so chosen that the gradient coupling prescription becomes exact in the soft-pion limit. In fact it is an easy exercise for a graduate student to show that the nucleon Born term in the gradient coupling theory not only contains the usual nucleon pole term in the dispersion-theoretic sense but also gives rise to a constant term in the $A^{(+)}$ amplitude whose value is precisely the Adler value $G^2_{\pi NN}/m_N$. At threshold this enormously large $A^{(+)}$ term cancels exactly with the enormously large nucleon pole contribution which, of course, is just in the B amplitude. This is why the symmetric part of s-wave scattering is essentially zero near threshold. I would like to emphasize that all this argument has nothing to do with Gell-Mann's current algebra.

Now, what about the antisymmetric part? Here we *do* need current algebra. Using the standard soft-pion technique, it is possible to derive that the antisymmetric part of the soft-pion scattering amplitude on target A is given by

$$\mathcal{M}^{(-)} = -(1/2c_\pi^2)(q + q')_\mu \varepsilon_{\alpha\beta\gamma}\langle A'|j_\mu^\gamma|A\rangle$$

This gives rise to the scattering length formula

$$a_T = -(\mu_\pi/4\pi c_\pi^2)\mathbf{T}_\pi \cdot \mathbf{T}_A$$

obtained by Tomozawa, Weinberg, and others. This is a powerful result. Perhaps an analogy with electromagnetism may be helpful here. When a very soft photon is scattered by a proton, the fact that the proton has 18 or more states that can be excited in γp collisions is totally irrelevant; the only important thing is the electric charge of the proton, which directly gives rise to the Thomson limit. Likewise, when a soft pion is scattered by a proton (or by any target for that matter), the only thing that matters is the fact that the proton has a half-unit of isospin; moreover, the coefficient of the scattering length formula is a universal constant that can be obtained directly in terms of the ratio of the π^+ lifetime and the $^{14}0$ lifetime. Here we appreciate in a clear and direct way the importance of conserved current and universality in understanding strong interaction phenomena.

By this time you probably have anticipated what I am going to say next. The current-algebra formula for pion scattering lengths has exactly the same $\mathbf{T}_\pi \cdot \mathbf{T}_A$ dependence as my 1960 formula based on ρ exchange. This similarity is not accidental; after all in a theory in which the isospin current is proportional to the ρ field

$$j_\mu{}^\alpha = (m_\rho{}^2/f_\rho)\rho_\mu{}^\alpha$$

the matrix element of $j_\mu{}^\alpha$ can be written as

$$\langle A'|j_\mu{}^\alpha|A\rangle = (m_\rho{}^2/f_\rho)\frac{\langle A'|J_\mu{}^{(\rho),\alpha}|A\rangle}{m_\rho{}^2 - t}$$

where $J_\mu{}^{(\rho)\alpha}$ stands for the source of the ρ meson. Using this, we can immediately interpret the current algebra expression for the scattering amplitude by saying that s-wave scattering is due to the propagation of the ρ meson in the t channel.

To summarize, the advances in PCAC and current algebra in the past few years essentially justify the 1960 conjecture that the low-energy scattering amplitude is due to the nucleon Born term computed according to the gradient coupling prescription plus the ρ exchange term computed in accordance with the universality principle. More recent attempts to construct an effective. (phenomenological) Lagrangian for the low-energy πN interaction (e.g., Schwinger, Zumino, and Wess, etc.) implicitly or explicitly exploit this point of view. We can now say with some confidence that the basic mysteries of the πN scattering lengths I mentioned in the beginning have finally been solved. It is rather sad that it took so many years for people to become convinced of such a simple prescription, but I guess we can't do anything about it.

Comments

GEOFFREY F. CHEW

University of California, Berkeley, California

I want to make a brief conjecture, motivated by the dualism that Dolen Horn, and Schmid[1] have pointed out (which I think is one of the most fascinating developments of the past year) concerning the possibility of representing certain phenomena alternatively as resonances or as peripheral phenomena. We have heard much about this duality and I assume you know what I am talking about. A slight extension (which I think follows logically if you believe in the duality idea) is that the famous Deck[2] effect may not conflict with the $A1$ being a resonance. You remember that the Deck effect depends on a doubly peripheral mechanism. Let me draw the picture to remind you (Diagram 1). You start with a π and a proton and

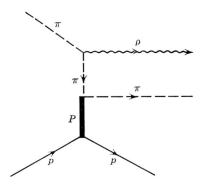

finally you have a ρ and a proton, but there is another π in the middle. According to Deck, the peripheral mechanism on the lower side is that of elastic scattering or as we Reggeists say, a Pomeranchuk P exchange. On the upper side it is a π exchange. If you are looking in the direction from left to right you can also say that the virtual π scatters elastically on the proton, but you can also say that the whole process is a doubly peripheral mechanism. Now, if you compute this as Deck did originally (it does not make much difference whether you Reggeize or not, because the π pole is so close to the physical region; Deck used an elastic amplitude for the lower part of the diagram, which is practically the same as using a flat P trajectory) what you find for the spectrum of the $\pi\rho$ system is a strong concentration near the threshold of that channel, and a big peak appears which looks somewhat similar to the observed peaks which have been associated with the $A1$.

This effect discovered by Deck is just one example of a very general mechanism, and it has undermined experimenters' confidence in their ability to establish resonances. Now, what I would like to suggest is that, following the Dolen, Horn, and Schmid idea, just because the doubly peripheral mechanism can generate a peaked concentration in the $\pi\rho$ spectrum does not mean that you have to reject the idea of a resonance, or several resonances. There may be two different, but both correct, ways of looking at the same phenomena. Of course, the spectrum when it is examined in detail, presumably would require the resonance picture to give an accurate representation, but the doubly peripheral approach may give a meaningful *average* representation of the same thing. So I would urge experimenters not to over-simplify this question and not to say " whenever I can find a peripheral mechanism which produces a peak, then I have to be especially careful about interpreting peaks as resonances." I would almost, taking a bootstrap point of view, put it the other way around: that if there exists a peripheral mechanism which produces a concentration at low energies, this may very well be just the situation where resonances will occur. It is essentially the same idea, I think, as the bootstrap mechanism which Mandelstam and Schmid have discussed in a singly peripheral context.

References

1. R. Dolen, D. Horn, and C. Schmid, *Phys. Rev. Letters*, **19**, 402 (1947).
2. R. Deck, *Phys. Rev. Letters*, **13**, 169 (1964).

Author Index

Numbers in parentheses are reference numbers and indicate that the author's work is referred to although his name is not mentioned. Numbers in *italics* show the pages on which the complete references appear.

A

Abers, E., 119(18), *130*, 134(3), *139*
Abrams, R., 187, *206*
Adelman, S. L., 47(1), *78*
Adelson-Volsky, G. M., 106(33a), *113*
Ademollo, M., 161(9), *165*
Adler, S. L., 155, 159, 160, 162(12), 163 (15), 164, *164, 165*
Alexander, G., 47(1), *78*
Alles-Borelli, V., 47(1), *78*
Almeida, S., 47(1), *78*
Altarelli, G., 145(22), *153*
Amati, D., 42(29), *45*, 162, *165*
Amblard, B., 104(29a), 105(29a), *112*
Anderson, E. W., 47(1), *78*
Anderson, H. L., 86(7), *112*
Anderson, J., 55, 69(26), 70(26), 72, 75 (26), *80*
Ankenbrandt, C. M., 47(1), *78*
Arbab, F., 142, 144(10), *153*
Armenteros, R., 69(30), 71(30), 75(27,30), 76(30), *80*
Arndt, R. A., 32(13), 38, 42(25), *44, 45*
Arnold, R. C., 62, *79*
Ashkin, J., 3(1), *24*
Ashmore, A., 106(31), *113*, 148(36), 149 (36), *154*
Astbury, A., 106(31), *113*
Atkinson, D., 33(14), *44*
Atkinson, H. H., 9(12), *25*
Auvil, P., 115(3), *129*
Azhgirei, L. S., 38(24), *45*

B

Baacke, J., 142(9), 144(9), *153*
Balachandran, A. P., 155(3), 160(3), *164*

Ball, J., 119(19), 120(12), 124(30), 126 (30), *130*
Barash-Schmidt, N., 189(8), 198(8), *206*
Barbaro-Galtieri, A., 73, 189(8), 198(8), *206*
Bareyre, P., 9(11), 14(19), 15(23), 17, 18 (19), 19, *25, 26*, 27(5), 29(5), *44*, 60, 66 (23), 67(23), 69(23), *79*, 115(4), 122(4), 126(4), 128(4), *129*
Barger, V., 27, 34(16), *44*, 135(6), 136, 138, *139*, 147(27), 150–152, *154*
Barloutaud, R., 56, 57, 69(30), 71(30), 75 (27,30), 76(30), *80*
Baroni, G., 106(31), 107, *113*
Bassel, R., 134, *139*
Batusov, 54
Beaupre, J., 145(15), *153*
Behrends, F. A., 39(28), *45*
Bell, 88
Bellamy, E. H., 4(2), 5(2), *24*
Belletini, G., 106, *113*
Bellettini, C., 47(1), *78*
Bennett, G., 134(1), *139*
Berge, 73
Berger, E. L., 47(3), *78*
Berley, 72
Bertanza, 55
Bethe, H. A., 88, *112*
Bigi, A., 104(29b), *112*
Birge, 57
Blair, I. M., 47(1), *78*
Blokhintsev, D. I., 103(28), *112*
Bogolubov, N. N., *112*
Bonamy, P., 145, *153*
Booth, N. E., 7(8), 9(13), *25*
Borgeoud, P., 104(29a), *112*
Borgese, A., 145(22), *153*
Borghini, M., 148(34), *154*

217

Durso, J. W., 42(29), *45*
Dyson, F., 117(15), *129*

E

Eandi, R. D., 69(25), *80*
Eden, R., *112*
Esterling, R. J., 7(8), 9(13), *25*

F

Faiman, D., 205, 206, *207*
Falk Variant, P., 104(29a), 105(29a), *112*
Falla, D. F., 106(31), *113*
Fayazuddin, 161, *165*
Feld, B. T., 17, 162(14), 163(14), *165*
Femino, 54
Ferro-Luzzi, M., 69(30), 71(30), 75(27, 30), 76(30), *80*
Feynman, R. P., 155(5), *164*
Filthuth, H., 69(30), 71(30), 75(27,30), 76 (30), *80*
Finkelstein, J., 145(19), *153*
Fischer, J., 4(4), *25*, 27(6), 36(19), *44*
Fischer, W., 145(14), *153*
Flatté, 73
Fletcher, R., 29(11), *44*
Focacci, M. N., 48(4), 49(4), *78*
Foley, K. J., 47(1), *78*, 87(14,15), 89(14, 21,22), 91(23), 93(14,15,21,22), 94(37), 95(22), 106(30,31), *112*, *113*, 146(26), 147(26), *154*
Franco, V., 134(2), *139*
Frautschi, S., 141, 147(30), *154*
Frazer, W., xiii, 35(17), *44*
Freedman, D. Z., 63(16), *79*, 168(3), 170, *185*
Froissart, M., 117, *129*
Frye, G., 117, *129*
Fubini, S., 33, *44*, 162, *165*
Fulco, J., xiii, 35(17), *44*, 117(11), *129*
Furuichi, S., 42(29), *45*

G

Gadjdicar, T. J., 100(25), *112*
Galbraith, 93
Garg, R., 118(16), 124(30), 126(30), *129*, *130*

Gellert, E., 47(1,3), *78*
Gell-Mann, M., 147, *154*, 145(23), *153*, 155, 156(6), *164*
Gensollen, 55, 56
Gerstein, I. S., 163(15), *165*
Giacomelli, G., 48(4), 49(4), *78*, 187(5,7), *206*
Gilman, F. J., 161(9), *165*, 184, *186*
Gilmore, R. S., 27(7), *44*, 87(14), 89(14), 93(14), 94(37), 106(31), *112*, *113*, 187 (6), *206*
Goldberg, H., 163(15), *165*
Goldberger, M. L., 82(1), 86(9), *111*, *112*
Goldhaber, G., 47(2), *78*
Gorn, W., 4(7), 6, *25*
Granet, P., 69(30), 71(30), 75(27,30), 76 (30), *80*
Grannis, P. D., 8(10), *25*
Green, H. S., 202, *207*
Greenberg, O. W., 190, 194, 197, 202, 205, 206, *206*, *207*
Gregorich, D., 145(23), *153*
Gribov, V. N., 168(1), *185*
Gross, D. J., 145(18), *153*
Gross, F., 163(15), *165*
Guillard, J. P., 104(29a), 105(29a), *112*
Guisan, O., 104(29a), 105(29a), *112*
Gundzik, M. G., 155(3), 160(3), *164*

H

Hambrecht, B., 155(3), 160(3), *164*
Hamilton, J., 38(23), 39, 43, *45*, 86(11), *112*, 117(13,14), 119(14), *129*, 162, 163, *165*
Han, M., 204, *207*
Hansroul, M., 8(10), *25*
Harari, H., 184, *186*
Harrington, 135(5), *139*
Hattersley, P. M., 27(7), *44*, 187(6), *206*
Heard, K. S., 4(6), 8(6), 9(12), *25*, 27(4), *44*
Heinz, R. M., 37(21), 39(21), 42(21), *44*, 135(8), *139*
Helmholz, A. C., 69(25), *80*
Hendry, A. W., 205, 206, *207*
Hepp, K., 83(5), *112*
Hepp, V., 69(30), 71(30), 75(27,30), 76 (30), *80*

Subject Index